AF723014

Radiation in Medicine and Biology

Radiation in Medicine and Biology

edited by
Pandit B. Vidyasagar
Sagar S. Jagtap
Omprakash Yemul

Pan Stanford Publishing

Published by

Pan Stanford Publishing Pte. Ltd.
Penthouse Level, Suntec Tower 3
8 Temasek Boulevard
Singapore 038988

Email: editorial@panstanford.com
Web: www.panstanford.com

British Library Cataloguing-in-Publication Data
A catalogue record for this book is available from the British Library.

Radiation in Medicine and Biology

Copyright © 2017 Pan Stanford Publishing Pte. Ltd.

All rights reserved. This book, or parts thereof, may not be reproduced in any form or by any means, electronic or mechanical, including photocopying, recording or any information storage and retrieval system now known or to be invented, without written permission from the publisher.

For photocopying of material in this volume, please pay a copying fee through the Copyright Clearance Center, Inc., 222 Rosewood Drive, Danvers, MA 01923, USA. In this case permission to photocopy is not required from the publisher.

ISBN 978-981-4745-92-5 (Hardcover)
ISBN 978-1-315-20656-1 (eBook)

Printed in the USA

Contents

Preface

Electromagnetic radiation includes radio waves, microwaves, visible light, ultraviolet light, and ionizing radiations such as X-rays and gamma rays. Although unplanned and accidental exposure to ionizing radiation enhances the risk of harm to living beings, therapeutic and diagnostic applications of ionizing radiation in medicine have tremendously contributed to improvement in human health. These medical applications include diagnostic radiology, radiation therapy for the treatment of cancer, and nuclear medicine. After the discovery in 1895 by W. Rontgen, X-rays were immediately applied to diagnostic radiography and therapy in medicine. Ionizing radiations have also been used for food preservation and genetic modification.

Cancer continues to kill millions of people worldwide. According to a World Health Organization report (2012), the number of new cancer cases worldwide is expected to rise by about 70% over the next two decades. More than 60% of the world's total new cancer cases and 70% of the world's cancer deaths occur in economically backward low- and middle-income countries. These countries have limited or least access to cancer treatment, particularly in high-technology-driven treatments such as radiation therapy. Radiation therapy is one of the most effective modalities of cancer treatment; approximately 50% of cancer patients receive radiation therapy in curative, adjuvant, or palliative settings during the course of their treatment.

Radiation dosimetry is one of the most crucial aspects of medical applications of ionizing radiation. Accuracy and precision of dosimetric quantification and measurement of absorption of radiation energy, i.e., absorbed dose, are vital in standardizing and developing therapeutic, diagnostic, experimental, research, and commercial applications of ionizing radiation. Radiation dosimetry of biological and clinical radiation therapy units has a direct impact on the accuracy and precision of radiation dose delivery and hence a significant impact on success of thetreatment.

The progress of biomedical sciences to the current level owes deeply to the constructive role played by physical and biophysical principles, techniques, and tools. In this era of rapidly evolving scientific knowledge and innovations, increasingly complex and intelligent technologies have created newer avenues for collaboration between the disciplines physical and medical sciences. Physical and biophysical methods have played an important role in the diagnostics and therapeutic strategies for the treatment of diseases and continue to create unique opportunities for multidisciplinary collaboration between medical and physical science researchers, including cancer research.

The Department of Physics, University of Pune, India, is one of the leading academic and research institutions in India. Traditionally, the department has pursued experimental and theoretical research in physics. In view of evolving multi-disciplinary trends discussed earlier, the focus was later diversified to interdisciplinary research and development problems pertaining to interfaces between physics, biology, and medicine. These evolving areas of research were further expanded to other fields, e.g., radiation biology, gravitational biology, and nanoparticle applications.

Prof. P. B. Vidyasagar's keen interest in interdisciplinary research and initiatives has created collaborations between academic researchers in the department and clinical physicists involved in cancer treatment at hospital-based radiation therapy facilities. Prof. Vidyasagar's initiatives led to PhD thesis works in important areas of medical physics, such as chemical radiation dosimetry, radiobiology, and Monte-Carlo simulation–based investigations of cobalt-60 tomotherapy, by his hospital-based medical physicist students. Other areas of research work in the department include diverse topics such as the study of low- and high-atomic-number targets for X-ray production, radio-electro chemotherapy, nanoparticle applications in radiation therapy, effects of altered gravity on various plant seedlings, and other important areas in biophysics. The interdisciplinary research and development work conducted by the physics department over the years has produced a significant body of work. This work highlights different applications of physics principles in medicine and biology and physics' contribution towards the improvement of human health, including the treatment of cancer.

The available body of work has inspired us to compile the material in this book. The book contains 10 articles and is categorized into two parts viz., new techniques in radiation therapy and effects of ionizing radiations on biological systems. The chapters are written by academicians, scientists, and clinical practitioners with wide experience in their fields. We believe this book will be immensely useful to students, teachers, academicians, and scientists with interest in interdisciplinary applications of radiation physics, including the areas of cancer therapy, nanomedicine, nanobiology, and gravitational biology.

We are thankful to all the eminent academicians and researchers who have authored valuable contributions for this book. We are grateful to Stanford Chong and Pan Stanford Publishing for encouragement and for agreeing to publish this book.

P. B. Vidyasagar
S. S. Jagtap
Omprakash Yemul

Foreword by Dr. Madan M. Rehani

The journey from intra-departmental compartmentalization to multidisciplinary interaction has been a long one that needs augmentation. While compartmentalization serves the purpose of in-depth specialization, the nature follows the principle of inter-dependence. Every species is dependent upon other for effective survival. Thus, neither complete dependence nor independence is what nature teaches us. We need to see how we can use excellence in one field for the benefit of other. Physics has generations of history and is crowned with many inventions, whereas medical sciences need help of basic medical sciences and paramedical and physical sciences to solve never-ending problems. Even after centuries, the need of medical sciences for support from physical and biomedical sciences will continue. It is in this respect that we need to remain cognizant that physicists should always feel that they are needed in medical sciences. Rather physicists should consider it as their responsibility to contribute to medical sciences. Every young graduate in physics should ask himself or herself what they are doing to alleviate the suffering of billions of humans, and it should create restlessness if there is no involvement in solving medical problems. No year should pass without any physics department highlighting what it has done to support medical sciences. There are countless ways by which physicists can contribute. After having spent nearly 44 years in medical physics, I cannot recall finishing any year when I felt that 365 days in the year were adequate for the work at hand. The whole ocean always lies unexplored, like the proverbial Newton's saying. Come out of your nutshells and the whole world is open to you. The current book goes in the direction of giving a small glimpse of what lies ahead for young physicists. The book deals with diverse areas as briefly noted below. The new advances in radiotherapy and imaging are showing much greater interaction to make radiotherapy more precise. Physicists are apt at precision and there cannot be any radiotherapy facility without a medical physicist. This becomes clear from the initial chapters in the current book. Investigations into cobalt-60 tomotherapy provide engaging actions for medical physicists as also FBX dosimetry for dose mapping. Fractionation is the heart and soul of radiotherapy and there is ever-increasing need for research. Incorporation of drugs into liposomes when combined with the application of electrical pulse provides challenging opportunities. Nanoparticles are currently gaining importance. Their synthesis and application in cancer imaging and therapy are discussed. Results of studies on change in membrane conductivity as a result of UV-C radiation are presented and finally attention has been drawn to the potential effects of the low level of non-ionizing radiation from mobile phones. It is natural for any new physicist to go through the cycle of enthusiasm, rational understanding, and critical assessment before being able to contribute. The availability of information in modern time is not lacking. What is becoming limiting is the human capacity and capability to grasp. One needs to be aware that there is no substitute for hard work of learning adequately in any area. I congratulate all involved in this book for taking a step that has the potential to show direction to young physicists what lies ahead, a vast ocean indeed, when they enter the arena of bio-medical sciences. Happy reading.

Dr. Madan M. Rehani

Massachusetts General Hospital, Harvard Medical School, USA
Director, Global Outreach on Radiation Protection
Vice President, International Organization for Medical Physics
Secretary, Protection in Medicine Committee
of International Commission on Radiological Protection (ICRP)
Adjunct Professor of Radiology and Medical Physics, Duke University, USA

Fighting Cancer by Radiation
Foreword by Prof. K. P. Mishra

I feel honored to be asked to write a foreword to the book *Radiation in Biology and Medicine*, edited by Pandit B. Vidyasagar and coeditors. Driven by my desire for seeking updates, I went through the volume with deep curiosity and found a comprehensive account of physical, chemical, and biological aspects of radiation therapy of cancer. Over the years, I have often felt the need of such a publication for the benefit of radiation researchers, clinicians, medical physicists, and instrument engineers. The book is a worthy source of reading, research, and reference.

Let me begin by recalling that the world of science witnessed unusual excitement with the chance discoveries of X-rays by W. R. Roentgen and radioactivity by Henri Becquerel at the end of 19th century. Despite revolutionary progress, the world is faced with increasing incidence of cancer, and each year 8–10 million patients succumb to cancer. In clinics, apart from surgery and chemotherapy, ionizing radiation–based radiotherapy is a frequently employed modality for treatment of cancer patients. Radiation kills cells, but it equally kills normal as well as tumor cells. Consequently, often, patient treatment by radiation has to be abandoned due to unacceptable adverse effects on their body. The challenge for the success of radiotherapy lies in designing the radiation source machines capable of delivering the dose targeted to diseased organ thereby minimizing exposure to adjacent cells. In addition, radiobiological research on cells, animals, and humans needs to identify molecular targets in tumor cells for their selective radiation killing while sparing the normal cells/tissues from harmful effects.

Early teletherapy machines used Cobalt-60 radiation source, which suffered from limitations of focused beam to the target, but constant technical refinement helped developing medical linacs offering isocenter dose delivery and flexibility of using electrons or X-rays for radiotherapy. This book gives an interesting account of technological progress from cobalt teletherapy machines to present-day medical linacs commonly employed in cancer therapy.

My irresistible interest in radiobiology and cancer research drives me to constantly watch progress especially in evaluating and analyzing available commercial radiation sources and newer radiobiological findings on cellular response to radiation. In accepting to write these lines for the book on radiation in biology and medicine, I am primarily guided by my urge to learn of recent advancements in the field. I must share that my curiosity was incredibly sustained in reading of the chapters from beginning to the end. Contributions have been meticulously organized making the volume a smooth and interesting reading for beginners as well as specialists in the field.

Without doubt, focused dose delivery to the tumor site is central to improving the quality of radiotherapy. In this regard, an impressive account has been provided on the generation of bremsstrahlung radiation from materials of low to high atomic number by S. D. Dhole and his colleagues using FLUKA simulation in optimizations of neutron dose level within acceptable limits. The issue of the quality of the radiation beam has further been extended by Joshi, Vidyasagar, and Schreiner, who utilized the Monte Carlo Code system to improve Co-60 source design for applications in tomotherapy. A systematic account has been given to determine potential radiation output enhancement for a fan beam with cylindrical and rectangular beam shapes. It is noteworthy that the suggested Co-60 source design with rectangular beam shape offers effective conformal dose delivery to patients in radiotherapy and tomotherapy. Radiobiological studies have revealed fractionation of total radiation dose to patients for maximizing the radiotherapy outcome by reducing adverse effects on normal tissues. However, challenge to basic radiobiologists and clinicians continues for the optimization of the treatment protocol in fractionated radiotherapy by reducing adverse radiation effects to healthy tissues but effectively killing tumor cells especially useful in the treatment outcome of squamous cell carcinoma for head and neck cancer patients.

It is pertinent to accurately quantify and control radiation doses given to patients in radiotherapy. I enjoyed refreshing my mind with a comprehensive account presented on the essential features of radiation dose measurement by ferrous sulfate-benzoic acid–xylenol orange (FBX) chemical dosimeter. The protocol is capable of determining radiation dose in diverse applications with photon- and electron-based radiotherapy. However, the need for further extensive research is much justified to develop a suitable dosimeter by overcoming the shortcomings of gel dosimetry.

More often, the treatment of cancer patients fails because cancer cells acquire resistance to radiation and therapeutic drugs. The mechanism of resistance is attributed to the non-internalization of drugs into target cells. A novel approach recently developed to deliver otherwise impermeant drugs into tumor cells employs the electroporation technology, which is based on a biophysical method of permeabilization of cell membrane. The research work of Pratip Sil and others in our group as well as the progress made in other laboratories around the world have made significant improvements in enhancing the radio and drug toxicity to tumor cells. A novel method of radio-electro-chemotherapy developed recently opens exciting new prospects for effectively dealing with untreatable cancers. It is hoped researchers would find it stimulating and engage themselves in optimizing this approach for practical clinical applications.

In past years, nanoscience and technology have opened exciting new vistas for the diagnosis and treatment of many diseases. To highlight growing new technologies, a chapter focuses on radiation-assisted synthesis of metal nanomaterials enabling the fine-tuning of defined molecular sizes and permitting multi-directional distribution of nanodrugs in the tumor volume. This novel synthetic protocol offers enormous new insights for improving tumor radio-toxicity. A potential formulation of gold nanoparticles is a frequently utilized approach in the radiotherapy of cancer, which has been adequately and lucidly described by Prabhakar Dongre of Mumbai University. However, the formulations require standardization and optimization for fruitful applications in clinic.

In view of the rapidly growing space expeditions, the area of space radiobiology has become a fascinating research direction with considerable implications to future space explorations. It is indeed interesting to understand the combined effects of altered-hyper-gravity and cosmic radiation on plants in controlled conditions. It is observed that the combined external agents cause deleterious effects on the growth of wheat seedlings and rendered the seed more sensitive to stress. This line of research is highly relevant to future space programs and intensified research is highly warranted. The effects of UV radiation on the properties of cellular membranes are of profound importance for evaluating cellular responses. The editors have appropriately included studies on UV-induced alterations in the properties of chitosan membrane. Studies have shown the influence of UV on alterations in the permeability of the cellular membrane. It is suggested that care be taken during the sterilization processes and practical use of the designed membrane.

The effects of mobile towers and mobile phone radiation on human health are controversial at present. The editors have thoughtfully included a chapter on this important topic. Because sample size is smaller and limited to adults, firm conclusions are difficult, but more intense research is highly suggestive. Some useful inputs have been provided for the reduction of radiation exposure by using cell phones only in emergency, curtailing the talk time, and communicating by text whenever possible.

I am immensely impressed with the wide range of frontline topics covered in this fascinating book devoted to cancer radiotherapy. Needless to say, the rich contents of the book will prove to be a good reference and will help catalyze future advances in the improvement of the radiation therapy of cancer patients. Readers will feel inspired and motivated to take up the challenges of problems related to human health.

Prof. K. P. Mishra, PhD
Present: Professor Emeritus, BM International Research Center
Jain Vishva Bharati Institute (Deemed University), Ladnun, Rajasthan, India
Former Senior Scientist and Head
Radiation Biology and Health Sciences Division
Bhabha Atomic Research Centre, Government of India
Mumbai, Maharashtra, India
Former Vice Chancellor
Nehru Gram Bharati University, Allahabad, Uttar Pradesh, India
Email: mishra_kaushala@rediffmail.com

PART 1

NEW TECHNIQUES IN RADIATION THERAPY

Chapter 1

Generation of Bremsstrahlung Radiation from Different Low- to High-Z Targets for Medical Applications: A Simulation Approach

Bhushankumar Jagnnath Patil,[a] Vasant Nagesh Bhoraskar,[b] and Sanjay Daga Dhole[b]

[a]*Department of Physics, Abasaheb Garware College, Karve Road, Pune, India*
[b]*Department of Physics, Savitribai Phule Pune University, Ganeshkhind, Pune, India*

sanjay@physics.unipune.ac.in

When high-energy electrons pass through a target (e–γ target), it generates a cascade shower of bremsstrahlung radiation with continuous energy spectrum shows an end point equal to the electron kinetic energy. A study of bremsstrahlung spectra from 6–18 MeV electron beam on different e–γ targets were carried out using FLUKA simulation. The study includes different materials as bremsstrahlung producing targets with variation of target thickness. The contribution of electrons, positrons and neutrons in the bremsstrahlung dose were calculated for each case and reported in the chapter. The data generated will help researchers and medical physicists to take precise right hand data of flux

Radiation in Medicine and Biology
Edited by Pandit B. Vidyasagar, Sagar S. Jagtap, and Omprakash Yemul
Copyright © 2017 Pan Stanford Publishing Pte. Ltd.
ISBN 978-981-4745-92-5 (Hardcover), 978-1-315-20656-1 (eBook)
www.panstanford.com

and energies of bremsstrahlung radiations to be used for desire object in radiation therapy. Moreover, for production of a clinical photon beam (radiotherapy application), the design of accelerator head assembly was optimized using 15 MeV electrons. The e–γ target, primary and secondary collimator, were designed in Monte Carlo based FLUKA code and the corresponding neutron dose equivalent and the gamma dose at the patient plane were estimated at various field sizes. The maximum neutron dose equivalent observed near the central axis of 30 × 30 cm^2 field and has a value 36.6 mSv/min. This is 0.61% of the central axis photon dose rate of 60 Gy/min. The values fall within the allowed limit by International Electrotechnical Commission (IEC). The dimensions of the collimators and filters were optimized in such a way that the neutron dose equivalent estimated is below the allowed limit in the therapy beam.

1.1 Introduction

1.1.1 Historical Background

In the late 19th century, Roentgen, Becquerel and Curie discovered the X-rays, natural radioactivity, and isolated radium, respectively [1]. However, X-rays were immediately applied to diagnosis and therapy in medicine. These discoveries paved the ways for radiation therapy. Five years after the discovery of radium, it was used for the first time to treat skin cancer. In 1913, Coolidge developed hot cathode X-ray tube, which enables external beam radiotherapy. Around 1920, brachytherapy was widely used to treat accessible tumors with radium needles or tubes [1]. To address the question of radioprotection International Commission on Radiological Protection (ICRP) was created in 1928. Moreover, at the same time the ionizing chamber and Geiger Muller chamber was also discovered to measure the accurate dose received by patients.

Later on, teletherapy unit was designed by using cobalt-60 source of high-energy gamma rays. The Co-60 gamma radiation unit was invented by H. E. Johns in Canada in the early 1950s for radiotherapy. The first telecobalt unit was installed in Hamilton, Canada, and widely used worldwide [1]. The only disadvantage while working with the radioactive sources was to manage the waste. This leads to the replacement of Co-60 teletherapy with a linear accelerator. During the first 50 years of radiotherapy, the technological progress was relatively slow and mainly based on X-ray tubes, Van de Graaff generators, and betatrons. The concurrently developed medical linacs, became the most widely used radiation source in modern radiotherapy. With its compact and efficient design, the linac offers excellent versatility for use in radiotherapy through isocentric mounting and provides either electron or megavoltage X-ray therapy with a wide range of energies [2–4]. The technology of radiation production has first been aimed to get higher photon and electron beam energies and intensities, and more recently towards computerization and intensity modulated beam delivery [5, 6].

1.1.2 Radiation Therapies for Cancer Diseases

Cells are the basic units of human body; in a normal body, cells grow, divide to make new cells, and die in an orderly way. Cancer begins when genetic changes impair this orderly process and it is because of changes to DNA (deoxyribonucleic acid) of a cell. In a normal cell, when DNA is damaged, the cell either repairs the damage or dies. In cancer cells, the damaged DNA is not repaired, but the cell does not die like it should. Instead, the cell goes on making new cells that the body does not need. These new cells all have the same damaged DNA as the first cell does. Such abnormal cell growth in any part of the body causes cancer. These cells may form a mass called a tumor. As a cancerous tumor grows, the bloodstream or lymphatic system may carry cancer cells to other parts of the body. For most cancers, a biopsy is the only way to make a definite diagnosis. Different types of cancer can behave very differently. For instance, lung cancer and skin cancer are very different diseases. They grow at different rates and respond to different treatments. Therefore, the correct treatment related to type of cancer is more important for cancer patient [7].

There are different ways of cancer treatment. The mostly used cancer treatments are surgery, radiation therapy, and chemotherapy. Surgery is a procedure in which a doctor with special training, called a surgeon, removes cancer from the patient's body. In radiation therapy, high radiation doses are used to kill cancer cells and shrink tumors. At high doses, radiation kills cancer cells or slows their growth. While in chemotherapy, drugs are usually used to kill cancer cells. Most patients have a combination of treatments, such as surgery with chemotherapy and/or radiation therapy. Radiation may be given before, during, or after surgery. The aim of radiation therapy to use before surgery is to shrink the size of the cancer. It may be used during surgery, so that it goes straight to the cancer without passing through the skin. Radiation therapy used this way is called intraoperative radiation. Sometimes radiation therapy is used after surgery, to kill any cancer cells that may remain or left in the surgical part of the body [7].

There are two main types of radiation therapy: One is an external beam and the other is located internally itself. External beam radiation therapy comes from a machine. Based on the tumor position and size, the type of radiation and the respective energy is decided. Internal radiation therapy is a treatment in which a source of radiation is put inside the patient's body. The radiation source can be solid or liquid. Internal radiation therapy with a solid source is called brachytherapy. It is used to treat cancers of the head and neck, breast, cervix, prostate, and eye. Liquid forms of internal radiation are most often used to treat thyroid cancer [7].

1.1.3 Today's Status in the World and India

Presently, there are more than ten thousand of accelerators running all over the world, out of which almost 50% are devoted to the medical applications. The main areas of use are radioisotopes production, radiography, and conventional radiotherapy with electron and photon beams [8]. Electrons and photons were found to be good members of radiation therapy for treating the cancer years ago. This is because of their high penetrability, low linear energy transfer (LET) to exhibit damage to the normal cell, and unique characteristics of dose distribution at depth [9, 10]. With the advent of high-energy linear and circular accelerators, electrons/photons have become a viable option in treating superficial tumors up to the depth of about 5–10 cm [11, 12].

External beam radiation therapy is also carried out with heavier particles such as neutrons produced by neutron generators and cyclotrons; protons produced by cyclotrons and heavy ions (helium, carbon, nitrogen, argon, neon) produced by Van de Graaff, synchrocyclotrons, pelletrons, and synchrotrons. In case of neutrons, the recoils and nuclear disintegration product contributing to the dose are responsible for a high-energy transfer to the biologically active molecules and destroy them in turn. High relative biological effectiveness (RBE), LET characteristics, and comparatively good dose distribution advantage are the main features of neutron therapy. As the biological effectiveness of neutrons is high, the required tumor dose is about one third the dose required with photons. Therefore, neutron therapy is presently realized in two versions: neutron capture therapy (NCT) and fast neutron therapy (FNT). In NCT, the isotope with large absorption cross section for thermal/epithermal neutrons is introduced into the body mainly through the blood, while FNT uses fast neutron with high penetrability and treats the malignant tumors of the head, neck, dairy gland, osteogenous sarcomas, etc. The limitations have been mainly due to difficulty in generating neutron particles as well as the construction of such treatment facilities [13, 14].

Proton beams are also a newer form of particle beam radiation used to treat cancer. They can offer better dose distribution due to its unique absorption profile in tissues, known as the Bragg peak, allowing the deposition of maximum destructive energy at the tumor site, while minimizing the damage to healthy tissues along their path. These have particular clinical use in pediatric tumors and in adults tumors located near critical structures such as spinal cord and skull base tumors, where maximal normal tissue sparing is crucial [15]. Particle radiation has higher LET than photons with higher biological effectiveness. Therefore, these forms of radiations may be more effective to the radio-resistant cancers such as sarcomas, renal cell carcinomas, melanomas, and glioblastoma [16]. However, the equipment for the production of particle radiation therapy is considerably more expensive than for photons. The decreasing costs of cyclotrons are likely to result in a wider use of proton beam therapy in the future [17].

1.1.4 Importance and Objective

Mega electron voltage (MeV) bremsstrahlung produced by medical accelerators is a common form of treatment modality for malignant tumors that occur at depth below the skin surface. The aim of radiation therapy is to deliver a sufficiently high absorbed dose to a defined target volume resulting in the eradication of the tumor with as minimal damage as possible to the surrounding healthy tissues [10]. Electron beams are used either as the primary mode of radiation therapy or combined with photon beams. Photon beams with energies higher than 10 MeV are preferred if doses should be delivered to larger depths (e.g., for the treatment of prostate cancer) and to enhance the skin sparing. In radiation therapy with external radiation beams, the absorbed dose at the specific point in a patient should be known with an overall uncertainty of 3.5% [14, 18, 19]. The main reason for the requirement of high accuracy in dose delivery is typically narrow margin between the dose needed for tumor control and the dose causing complications for healthy tissues. Besides an individual treatment, the high accuracy in dose delivery is essential to enable a reliable analysis and a comparison of results of different radiation therapy techniques and modalities [19]. Therefore, the knowledge of the energy spectra and angular distribution of photon beams at the point of application is essential for accurate dose calculations [20]. The bremsstrahlung energy spectrum in the central part of a photon beam is somewhat harder than in the region near edge of the beam. Consequently, the relative depth dose in tissue will vary as a function of distance from the beam axis. Moreover, variation of spectrum across the beam is also important when calculating transmission through a block or a compensating device [20]. Knowledge of the angular distribution of photons is useful when calculating dose near the beam boundaries.

The emission of bremsstrahlung from a target also generates neutron if the incident energy goes beyond 10 MeV. The biological effectiveness of neutrons is substantially higher than that of photons [21]; therefore even a small neutron dose will increase the risk for secondary cancer. Therefore, it is necessary to know the contamination of positron and neutron in bremsstrahlung beam to minimize unwanted dose to the patient's body. Moreover, the clinically applicable photon beam is produced in an e–γ target, flattened with a flattening filter, collimated in primary collimator, and beam shaping done using secondary collimator. It is necessary to minimize the contribution of neutrons while designing such accelerator head assembly. For this work, FLUKA simulations have been carried out to evaluate the photoneutron yield and spectra produced through accelerator head assembly of 15 MeV medical linac as a function of the radiation field sizes. Therefore, in the present chapter, considering the importance of bremsstrahlung radiations in the medical field, the actual evaluation and designing aspect of the associated accessories of the sources have been studied thoroughly.

1.2 Interaction of Radiations with Matter

1.2.1 Interaction of Electrons with Matter

In an absorbing material, an electron is slowed down and finally brought to rest by the combined action of all four of these elastic and inelastic processes.

1.2.1.1 Elastic collision with atomic electron

An incident charged particle may be elastically deflected in the field of the atomic electrons of the struck atom. Such collisions are significant only for the case of very-low-energy (<100 eV) incident electrons.

1.2.1.2 Elastic collision with atomic nuclei

The incident particle is elastically deflected in the field of nucleus. This is a radiative free process. The deflection due to nucleus will take place at an angle greater than 90°. Cross section for this process is very small.

1.2.1.3 Inelastic collision with atomic electron

The incident electrons lose their energy while passing through matter. During such inelastic collisions, the incident electron transfers part of its energy to a bound atomic electron taking it to an unbound state (ionization) or an excited state (excitation). The cross section is proportional to Z of target and E of incident electron beam.

1.2.1.4 Inelastic collision with atomic nuclei

The incident electron passing through the field of a nucleus experiences a deflection with a resultant emission of radiation. This process is known as bremsstrahlung. This leads to loss of kinetic energy of the incident electron in the form of radiation loss. The rate of energy loss by this interaction is proportional to Z^2, where Z is the atomic number of the target atom [22].

The collisional energy loss is predominant for light elements and at low electron energies while radiative losses start becoming comparable only at high electron energies and for heavy targets. When a charged particle is either accelerated or decelerated in an electric field, electromagnetic radiations may be given off. If an electron passes close to nucleus while traversing a substance, the charge Z on the nucleus will exert a force on the electron. This will cause its path to be bent. During this acceleration, the electron may radiate energy of any amount from zero up to its total kinetic energy (E_{k}) in the form of bremsstrahlung. The total bremsstrahlung per atom is roughly proportional to $(Z/m)^2$, where Z is the atomic number of the absorbing matter and m is the mass of the charged particle. Because of a $1/\mathrm{m}^2$ dependence, the amount of bremsstrahlung is almost completely negligible for all particles except electrons, unless the particle energy is in the GeV range. As a electron traverses matter, it suffers many "soft" or "glancing" collisions with the atoms along its path. At each collision the particle loses energy and changes its direction slightly. The net result is that the electrons path is very tortuous. The range of the particle (R) ($\mathrm{gm/cm^2}$) is defined as the minimum thickness required of an absorber to stop the particle (electron).

1.2.2 Interaction of Photons with Matter

Different interactions dominate for different photon energies. The main modes of interaction of gamma rays with matter are the photo effect in its both photoelectric and photonuclear forms, Compton scattering, and electron positron pair production. To a minor extent, photofission, Rayleigh scattering, and Thomson scattering also occur.

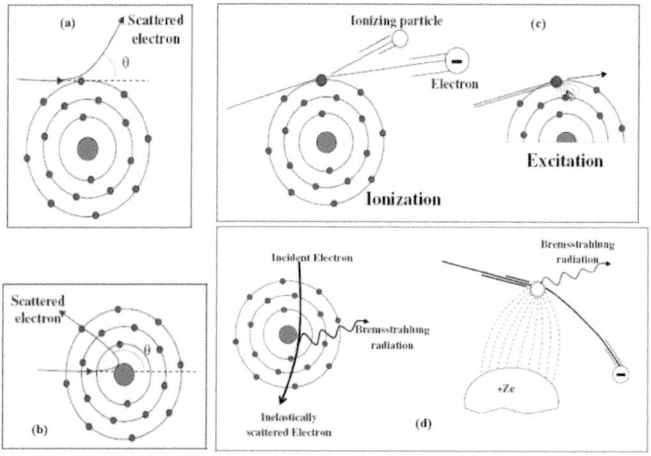

Figure 1.1 Interaction of electron with the matter.

1.2.2.1 Photoelectric effect

In this process, the photon is absorbed by an atom and expels an electron by losing all its energy in one interaction. The probability of photoelectric absorption is inversely proportional to the gamma photon energy and proportional to atomic number of the atom (Z^5). The probability is greater for the more tightly electron. The kinetic energy E_e of the emitted photo electron is given by ($h\gamma - E_\gamma$) [23].

1.2.2.2 Compton scattering

In this process, the gamma ray interacts with a free or weakly bound electron and transfers part of its energy to electron. The photon is scattered through an angle θ with an energy E_γ while the electron recoils with kinetic energy (E_e) at an angle ϕ. The kinetic energy of the electron is equal to the difference of the energy lost by the gamma ray and the electron binding energy: $E_e = h\gamma - E_\gamma$. The probability for this process is weakly dependent on E and Z. The interaction probability depends on the electron density, which is proportional to Z/A and nearly constant for all materials.

1.2.2.3 Pair production

In this process, the gamma ray loses all its energy in one interaction. A gamma ray with an energy of at least 1.022 MeV can create an electron–positron pair when it is under the influence of the strong electromagnetic field in the vicinity of a nucleus. A photon cannot create an electron–positron pair in free space, as the process cannot conserve momentum and energy. In this interaction, the nucleus receives a very small amount of recoil energy to conserve momentum, but the nucleus is otherwise unchanged and the gamma ray disappears. A heavier nucleus takes less recoil energy. This interaction has a threshold of 1.022 MeV because that is the minimum energy required to create the electron and positron. If the gamma ray energy exceeds 1.022 MeV, the excess energy is shared between the electron and positron as kinetic energy. Above the threshold, the probability of the interaction increases rapidly with energy. The probability of pair production is proportional to the square of the atomic number Z and is significant in high-Z elements [23].

1.2.2.4 Photonuclear absorption

It is a form of photoelectric effect where the photon is absorbed in a nucleus. When the energy of the photon is higher than the binding energy of the least bound neutron, photo neutron emission becomes possible. The reaming energy is shared as kinetic energy between the neutron and the residual nucleus. The photoneutron production threshold energy varies in general from 8 to 19 MeV for light nuclei ($A < 40$) and 6–8 MeV for heavy nuclei [24]. However, for a few light elements such as deuterium and beryllium, threshold energy is 2.226 and 1.666 MeV, respectively [25]. The cross section of this process is generally small, but peaks in the region of the nuclear "giant resonance." Neutrons are produced through following different mechanisms based on the incident photon energy as, giant dipole resonance ($E_{th}(n) < E <$ 30 MeV), Quasi-Deuteron Region (30 MeV $< E <$ 200 MeV), Delta Resonance ($E >$ 140 MeV) and vector meson dominance (in high-energy region above delta resonance). The boundaries between the four energy domains are not sharp and depend somewhat on atomic number.

1.2.3 Basic Mechanism of Cell Killing by Radiation

The biological target of radiation in the cell is DNA. Radiation can directly interact with cellular DNA and cause damage. In such case, the dose of radiation absorbed correlates directly with the energy of the beam and its deposition of energy in tissues, which results in damage to DNA strands and diminishes the cell's ability to replicate indefinitely. Moreover, the indirect DNA damage can cause by the free radicals, derived from the ionization or excitation of the water component of the cells. Figure 1.3 shows the cell killing mechanism in the body.

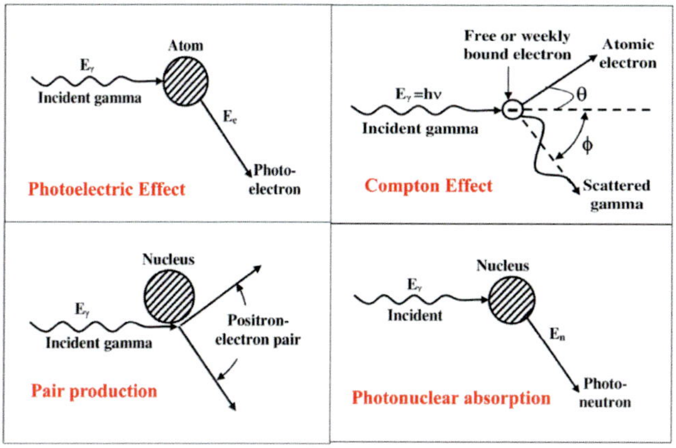

Figure 1.2 Interaction of photons/gamma with the matter.

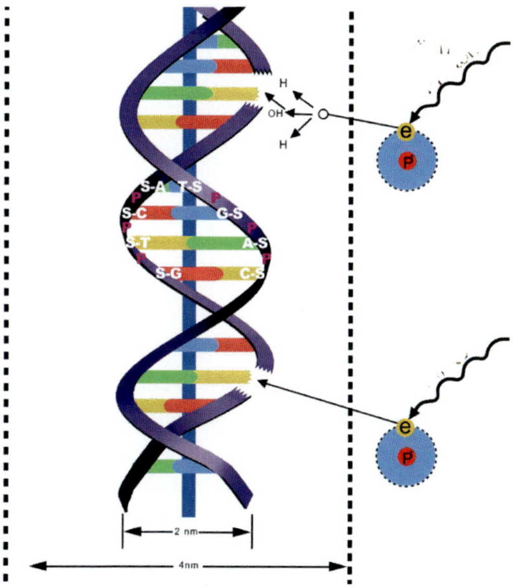

Figure 1.3 Cell killing mechanism.

The primary effect of radiation is the ionization of the water molecule,

$$H_2O \xrightarrow{\text{Radiation}} H_2O^+ + e^-$$

$$H_2O^+ + e^- \rightarrow H_2O^-$$

These molecular ions further break down

$$H_2O^+ \rightarrow H^+ + OH$$

$$H_2O^- \rightarrow H + OH^-$$

Thus, although the H^+ and OH^- ions will recombine to form water molecule, highly reactive free radicals H and OH still remain and can react with the molecule of the cell nucleus. If R–H is such a molecule then we get

$$R-H+OH \rightarrow R+H_2O$$

The free organic radical can also be formed by direct action of the radiation, i.e.,

$$R-H \xrightarrow{\text{Radiation}} R+H^+ + e^-$$

The presence of oxygen near the site of interaction may give rise to hydrogen par oxide (H_2O_2) or the peroxide radical, i.e.,

$$O_2 + H + H \rightarrow H_2O_2$$

$$R + O_2 \rightarrow RO_2$$

The creation of H_2O_2 and RO_2 leads to the damage of the cell.

Radiation either directly or indirectly damages the DNA strands, which results in single-strand breaks or double strand breaks (DSBs). Double-strand DNA breaks are irreparable and more responsible than the single-strand DNA breaks for most of cell killing in cancer as well as surrounding normal cells. Extensive damage to cancer cells DNA can lead to cell death. DNA double-strand breaks are more responsible for most cells killing, even a single DSB is sufficient to kill a cell or disturb its genomic integrity by the radiation treatment. For low LET radiations, the damage is induced primarily by activated radicals produced from atomic interactions. Over the energy range of therapeutically used X-rays, typically 100 keV to 25 MeV, approximately the same physical dose needs to be delivered at different energies to reach a given biological endpoint, resulting in similar RBEs. High LET radiations such as protons, neutrons, however, result in biological damage that is generally larger per unit dose than for X-rays, resulting in an elevated RBE [26].

1.3 Methodology

1.3.1 Monte Carlo-Based FLUKA Simulation

The radiation field of electron accelerator includes several components such as bremsstrahlung photons, fast neutrons, positrons, hadrons, and muons. The production and transport of all these radiations through different targets are difficult to study theoretically even on the basis of correct experiments. Therefore, simulations with an effective Monte Carlo code are very helpful to get information of all the particles produced in accelerator head and their dose deposition in patient's body. A general purpose Monte Carlo based code FLUKA [27] has been used for the calculations of particle transport and interactions with matter. FLUKA code version 2006 and 2008 has been used to calculate the results. It can simulate of about 60 different particles with high accuracy, including photons, electrons, neutrons, heavy ions, and antiparticles. The lowest transport limit for all particles is ~1 keV. There are various tools for input geometry visualization and output plotting in two and three dimensions giving a clear picture of the calculations. FLUKA can handle very complex geometries, using an improved version of the well-known combinatorial geometry (CG) package. FLUKA uses an original transport algorithm for charged particles, including complete multiple Coulomb scattering treatment. It also uses Bethe–Bloch theory for energy loss mechanism. Delta-ray production via Bhabha and Moller scattering is implemented in FLUKA. In FLUKA the full set of Seltzer and Berger cross sections [28] of accurate electron-nucleus and electron-electron bremsstrahlung has been tabulated in an extended form [29]. FLUKA scores fluence and current as a function of energy and angle. It can also score track-length fluence in a binning structure (Cartesian or cylindrical) independent of geometry.

The radiation units used in FLUKA to measure the number of particle generated through interaction mechanism are Flux (particles – cm^{-2} – sec^{-1} per primary particle), Particle Fluence (ϕ) (particles – MeV^{-1} – cm^{-2} per primary particle) and Particle Current (J) (particles – MeV^{-1} –cm^{-2} per primary particle). The particle current is defined as the number of particles crossing an area over a certain time period and represents the number of particles per unit area that passed during this time. Moreover, it also uses regular units like dose (J/kg, GeV/g, RAD, gray), dose equivalent (Sievert (Sv), Roentgen Equivalent Man (REM)) to estimate the energy deposited in the medium.

1.4 Case Studies

1.4.1 Case I: Design of e–γ Target for Radiation Therapy

The set of simulation was carried out to estimate the bremsstrahlung yield for

(i) Electron Energy: 6, 9, 12, 15, and 18 MeV
(ii) Materials used as e–γ target (Z): Be(4), Al(13), Si(14), Fe(26), Cu(29), Mo(42), Ag(47), Gd(64), Ta(73), W(74), Au(79), Pb(82), U(92). (The materials having higher melting point were chosen for bremsstrahlung study.)
(iii) Target thickness: ranging from 0.1% to 150% of range of electron in the material.

The inputs for FLUKA code are defined as the beam diameter 0.3 cm, beam direction along the Z axis and beam allowed to fall on the target perpendicularly. The cylindrical geometry was chosen and thickness of the cylinder was varied. An energy cut was set to be 100 keV for electron (Ecut) and photon (Pcut). Moreover, the 1×10^8 to 1×10^9 primary particles were used to run the system to minimize the statistical errors in the simulation. The estimated errors were found to be less than 0.1% in the bremsstrahlung case; it almost takes ~15–18 h to complete the run on Xeon quad-core processor.

The geometry used for the measurement of angular distribution of integrated bremsstrahlung fluence is shown in Fig. 1.4. An electron beam is incident on the e–γ target along Z direction and the generated radiations were estimated. Sphere of radius 100 cm has been divided by parallel planes along Z direction such that it can form 0.2 cm thick rings of sphere at various angles of 0 to 180° with respect to the incident electron beam. The horizontal lines in Fig. 1.4 correspond to the 0.2 cm thick rings for the measurement of particle fluence at various angles. The e–γ conversion performances of several materials have been accurately studied by simulating the bremsstrahlung spectrum obtained for each material as a function of the converter thickness for different electron energies.

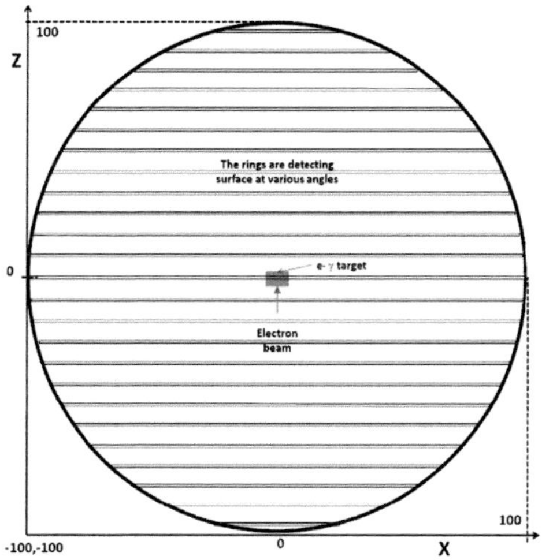

Figure 1.4 A view of XZ plane geometry used for the measurement of angular distribution of bremsstrahlung radiation in FLUKA simulation.

When high-energy electron interacts with material, it produces bremsstrahlung radiation through the inelastic collision of electrons with nucleus. The variation in integrated bremsstrahlung fluence with e–γ target thickness for 6 MeV incident electron energy is shown in Fig. 1.5. From the figure, it is observed that as the e–γ target thickness increases, the bremsstrahlung fluence also increases until certain thickness and then decreases with increase in the thickness of e–γ target. It is observed from the figure that the bremsstrahlung fluence peak is shifted towards the lower thickness of the target with increase in the Z of the target. For a higher-density material, the peak is found at lower thickness. This mainly attributes to the absorption of photon in the high density material itself. Moreover, the thickness that gives maximum bremsstrahlung fluence for e–γ target increases with the Z of the target and also with incident electron energy. The same trend is observed for the 9, 12, 15, and 18 MeV electron energy cases. It is also observed that uranium gives maximum bremsstrahlung fluence at the 0.08, 1, 1.2, 1.4 and 1.5 cm thickness of target for 6, 9, 12, 15, and 18 MeV incident electron energy, respectively, and beryllium gives lowest bremsstrahlung fluence for all energies.

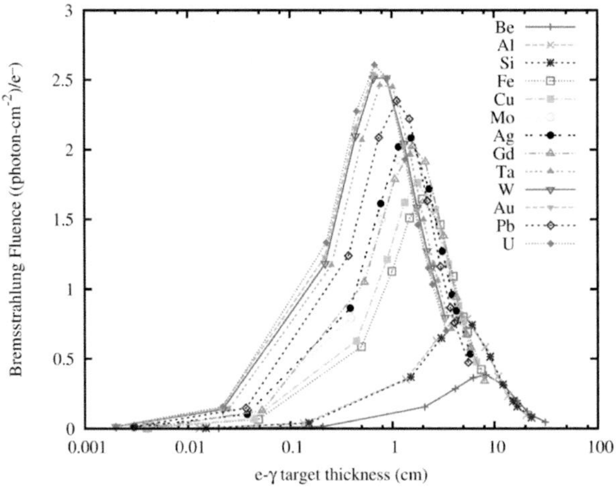

Figure 1.5 Integrated bremsstrahlung fluence versus e–γ target thickness for 6 MeV energy electron incident on different materials.

Typical photon fluence distribution in the XY plane (7 cm × 7 cm, 1 mm/bin) is shown in Fig. 1.6 for 6 MeV electron incident on 0.08 cm thick tungsten target. Figure 1.6 shows that the maximum photon fluence is in the forward direction, which is at the center and it drops radially.

The angular distribution of the relative bremsstrahlung fluence for each material at thickness equal to the range of respective energy of electron are shown in Fig. 1.7 for 6–18 MeV electrons. It is observed from figure that the relative bremsstrahlung radiations produced in the forward direction, i.e., at 0° is more and decreases exponentially with increase in the angle. In addition, the relative bremsstrahlung fluence decays very fast for low-Z materials and slowly for high-Z materials with increase in angle. However, until 8° to 9°, the fluence converges for all the materials. This is because dE/dx for electrons in the initial few angstrom layers of the target is same and further increases in angle; the electrons see long path and have observed more statistical fluctuations in the dE/dx. A similar trend is observed for the higher energies too.

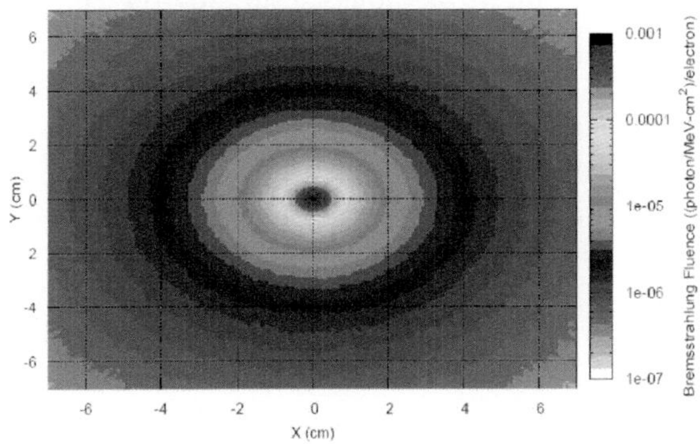

Figure 1.6 Photon fluence distribution in *XY* plane (7 cm × 7 cm, 1 mm/bin) for 6 MeV electron incident on 0.08 cm-thick tungsten target.

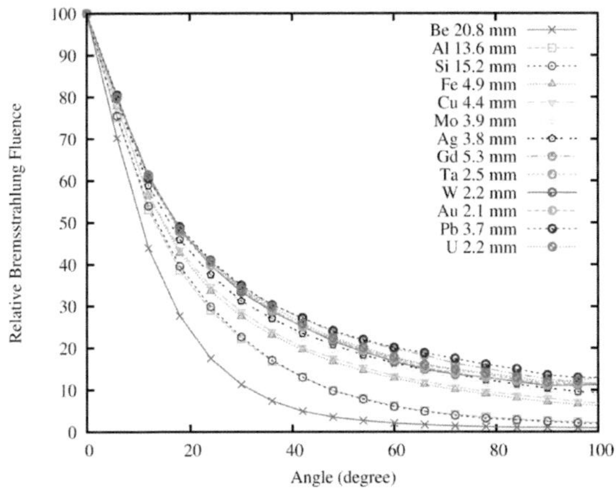

Figure 1.7 Angular distribution of bremsstrahlung fluence for 6 MeV energy electron incident on e-γ target of thickness equal to the range of electrons.

Figure 1.8 shows the variation in bremsstrahlung fluence with atomic number of e–γ target for 6 to 18 MeV electron energies. It is observed from the figure that the usually bremsstrahlung fluence increases with Z of the e–γ target except gadolinium and lead, where bremsstrahlung fluence is found to be less. This is attributed mainly to the fall in the density of the respective material. Variation in mean energy of bremsstrahlung radiation with atomic number of e–γ target is shown in Fig. 1.9. From this figure, it is seen that for 6 MeV energy electron the mean energy increases continuously with Z, whereas for 18 MeV incident electron there is fluctuation in the mean energy of bremsstrahlung radiation for Al, Si, Ta, W, and Au due to their respective variations in the absorption cross section for bremsstrahlung radiation. Variation in the contribution of positron fluence with atomic number of e–γ target is shown in Fig. 1.10. From this figure, it is also observed that the positron fluence increases with atomic number except Gd and Pb for which it is found less. This is because of the lower density for the respective target compared with others. Variation in the neutron fluence with atomic number of e–γ target for 6–18 MeV electron energy is shown in Fig. 1.11. It is observed from the figure that the neutron fluence increases with atomic number and mainly depends on the respective cross section of the targets.

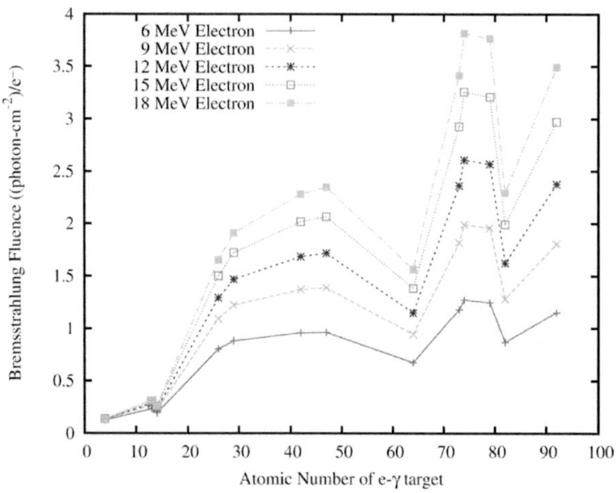

Figure 1.8 Variation in bremsstrahlung fluence with respect to the atomic number of e–γ target for 6–18 MeV incident electron energies.

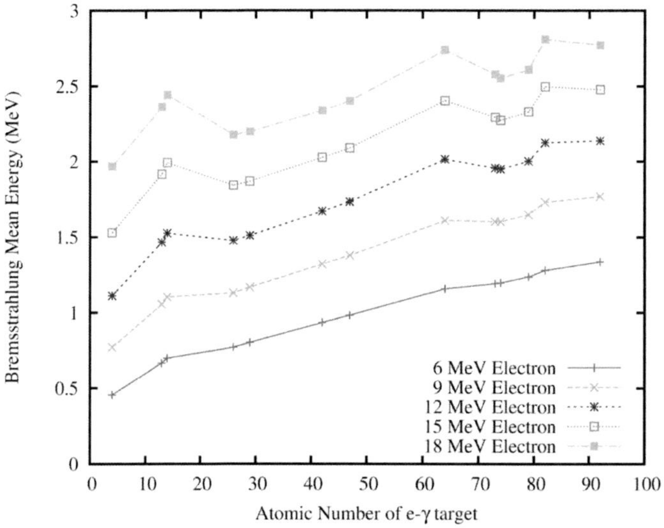

Figure 1.9 Variation in mean energy of bremsstrahlung with respect to the atomic number of e–γ target for 6–18 MeV incident electron energies.

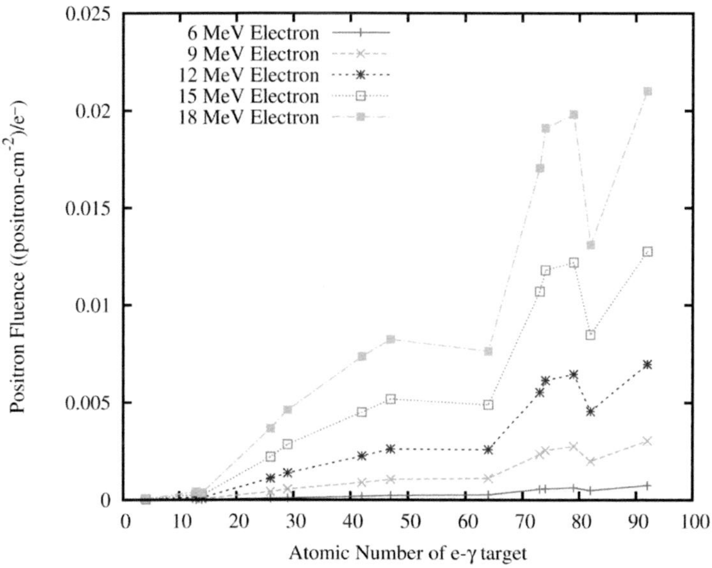

Figure 1.10 Variation in positron fluence with respect to the atomic number of e–γ target for 6–18 MeV incident electron energies.

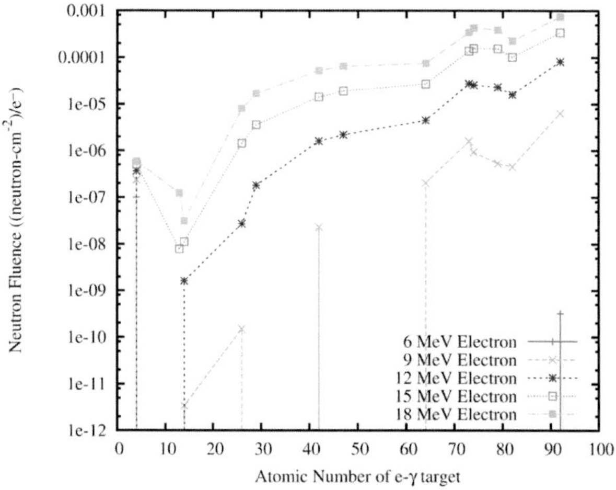

Figure 1.11 Variation in neutron fluence with respect to the atomic number of e–γ target for 6–18 MeV incident electron energies.

1.4.2 Case II: An Optimization of Accelerator Head Assembly for Radiotherapy

One of the major applications of the bremsstrahlung radiation is for radiotherapy. Photon beams with energies higher than 10 MeV are preferred when doses need to be delivered to greater depths (e.g., for the treatment of prostate cancer) and to enhance the skin sparing. The requirement for the production of clinical photon beam using high-energy electron is that the photon beam should have a spatially uniform fluence and well collimated in a reference plane that is perpendicular to the beam axis. Generally, this plane is defined at a depth of 10 cm in a water phantom. The surface of water phantom is 100 cm away from the photon source, i.e., the surface–source distance (SSD) is 100 cm. When the required condition is met, the radiation beam will produce a uniform dose distribution across the reference plane.

In case of high-energy photon a parasitic effect that occurs is the production of neutrons, mainly due to the photonuclear giant-dipole-resonance (GDR) reaction (γ; n) induced by high-energy photons in the accelerator head materials [30]. As the quality factor of the neutrons is higher as compared to the photons, the biological effectiveness of neutron is higher. Hence small neutron dose is also responsible for more damage to body and also it is on high risk for cancer cause [21]. To reduce the risk, one must have to take care of neutron fluence at the time of designing the accelerator head assembly. Moreover, the neutron energy spectra generated in the accelerator head assembly is also important for various purposes such as to optimize the room shielding and to estimate energy dependent neutron doses received by the patient and medical personnel working in and around the therapy facilities. Therefore, in the present work an accelerator head assembly has been designed for 15 MeV linear accelerator. To estimate the neutron contamination in terms of dose equivalent and energy spectra has been carried out by Monte Carlo based FLUKA simulations. It is also used to evaluate the photoneutron yield and spectra produced through accelerator head assembly of 15 MeV medical linac as a function of the radiation field sizes.

1.4.2.1 Accelerator head assembly

Clinical photon beams emanating from a medical linac are produced in the e–γ target, flattened with a flattening filter and collimates in collimators. The components of accelerator head assembly are shown in Fig. 1.12. The photon collimation and beam shaping is achieved with primary collimator and movable secondary collimator.

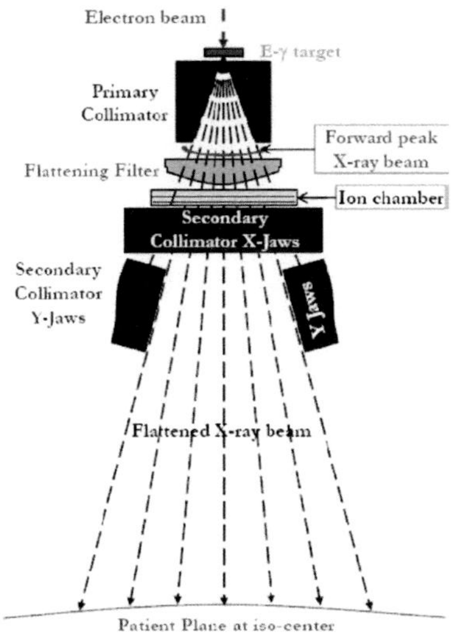

Figure 1.12 The schematic of various components of accelerator head assembly.

As the electron beam penetrated the target material, the Coulomb interactions resulted in the production of a photon beam, called Bremsstrahlung photon. The thickness of the e–γ target was optimized such that the entire electron incident on the target gets absorbed in target itself. Therefore, the target thickness was optimized more than the range of electron in e–γ target. Immediately after e–γ target is the primary collimator. It is a circular conical hole that defines the maximum divergence of the beam and therefore the maximum circular field size. For obtaining maximum field of 51 cm diameter at patient plane, the primary conical collimator comprises a 28° cone bored into a metal block. The distance between the e–γ target and primary collimator is optimized such that the system must produce the minimum penumbra. The thickness of the shielding block is usually designed to attenuate the mean primary X-ray beam intensity to be less than 0.1% of the initial value (three tenth-value layers (TVLs)). According to IEC recommendations, the maximum leakage should not exceed 0.2% of the open beam value. The high-Z element is often used for this component because of high attenuation coefficient. The photon beam exiting from the primary collimator does not have uniform spatial intensity. It has an angular distribution that is strongly peaked in forward direction with respect to the initial electron beam [31, 32]. A more uniform angular distribution of the photon beam can be achieved by passing it through a flattening filter. It is Gaussian shaped. The dose distribution is very sensitive to the position of the flattening filter. A small misalignment of the flattening filter within a few millimeters in the linac head would cause large variations in the dose distribution [33]. The flattening filter also has another effect on the beam called beam hardening [34]. The flattening filter not only hardens the beam as a whole but further enhances the relative hardness near the center. Following the flattening filter is the monitoring system to measure the radiation beam dose output. The mirror is used to project light from the optical source to replicate the shape of the radiation field. Below the mirror are two sets of jaws that constitute a secondary collimator. The maximum circular field defined by the primary collimator is truncated with an adjustable rectangular collimator, which consists of upper and lower independent movable jaws for producing rectangular and square fields with a maximum dimension of

40×40 cm^2 at the linac isocenter. These blocks have sufficient thickness to shield out unwanted radiations. These jaws are designed to move in an arc shape to produce different field size of the photon beam. The IEC recommends that the transmission of the primary photon beam through the rectangular collimator should not exceed 2% of the open beam value. The material and dimension of collimators were optimized such that the neutron contamination in the gamma beam was below the allowed limit.

1.4.2.2 Optimization of accelerator head assembly

Among the materials studied, tungsten was found to be best suitable as e–γ target of thickness 0.42 cm since it absorbs almost all the incident electrons. The mean energy of the bremsstrahlung spectrum produce by the e–γ target is around 2.033 MeV. Using the FLUKA simulations the tenth value layer (TVL) values of photon for iron, lead, tungsten, bismuth, tungsten + copper, and tantalum are estimated at 7.64, 3.87, 2.93, 4.40, 3.37 and 3.30 cm, respectively. The material that has less TVL thickness and low neutron production is the best material to be used for the primary collimator. Therefore, the W–Cu material has been optimized to design the primary collimator. In addition, the leakage radiations were calculated offside at a distance of 1 m from the beam center and found to be less than 0.2% of beam value (recommended by the IEC) for the thickness more than 8 cm of primary collimator. Therefore, it was optimized to use 10 cm thickness of W–Cu for primary collimator. The neutron fluence calculated at isocenter is 3.94×10^{-9} n–c m^{-2}/e$^-$.

A Gaussian-shaped flattening filter was divided into eight truncated right-angled cones (TRCs) of various thicknesses and of increasing radii. As the less number of neutron are produced from iron material compared with lead material, iron has been used as a filter. The dimensions of optimized flattening filter made of eight TRCs are shown in Fig. 1.13a. The flattened dose estimated in the water phantom is shown in Fig. 1.13b, and it gives flattened dose for 40×40 cm^2 field size.

The jaws of the secondary collimator were positioned such that the rotation of respective X jaws and Y jaws forms a square field size. The thickness of secondary collimator was optimized such that the transmission of the primary X-ray beam does not exceed 2% of the open beam value. Therefore, the thickness of the secondary collimator was optimized to 8 cm. The rotation of the X and Y jaws of secondary collimator along the arc changes the radiation field size area from 0 × 0 to 40 × 40 cm^2. Using calculated positions of jaws for each field size, the accelerator head assembly has been modeled in FLUKA to estimate the gamma absorbed dose and neutron dose equivalent. In addition, the neutron fluence and respective spectra for different field sizes have also been estimated.

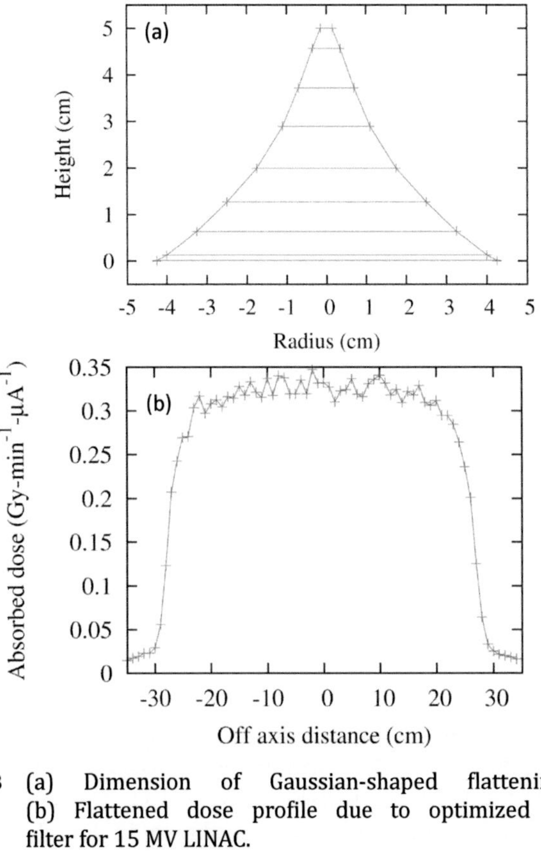

Figure 1.13 (a) Dimension of Gaussian-shaped flattening filter. (b) Flattened dose profile due to optimized flattening filter for 15 MV LINAC.

Figure 1.14 shows the relative photon depth dose distribution, for 10 × 10 cm^2, 20 × 20 cm^2, 30 × 30 cm^2, and 40 × 40 cm^2 field sizes at an SSD of 100 cm using the optimized accelerator head geometry. The distance at which the maximum dose is delivered in water is 2 cm.

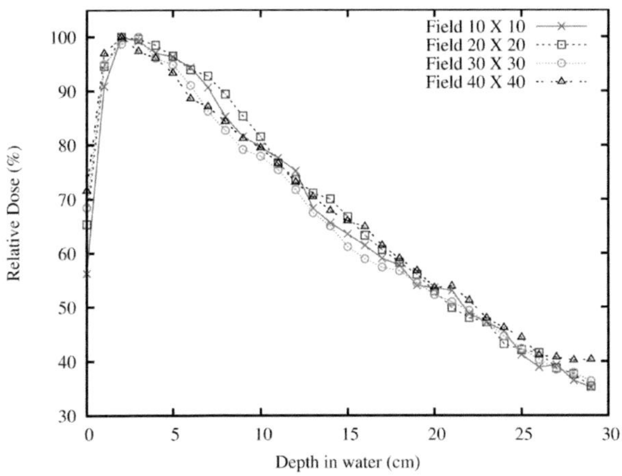

Figure 1.14 Relative photon depth dose distribution for various field sizes at SSD of 100 cm.

Figures 1.15a and 1.15b show the flattened and unflattened dose profiles in water phantom for different field sizes. The ratio of the neutron dose equivalent to the central axis photon absorbed dose along the longitudinal axis at patient plane for different field sizes is shown in Fig. 1.16. The maximum neutron dose equivalent is observed near the central axis of 30 × 30 cm^2 field size. This is 0.7 1% of the central axis photon dose rate of 0.3 Gy/min at 1 μA electron beam current. The values of the neutron dose equivalent estimated are consistent with the results of other measurements reported in literature [35] and fall within the allowed limit by the IEC. In addition, the ratio of neutron dose equivalent to central axis photon dose was maintained below the allowed limit set by the IEC (<1 mSv/Gy) inside and 0.5 mSv/Gy outside of photon field. Table 1.1 shows the bremsstrahlung fluence, maximum bremsstrahlung dose delivered in water, and neuron fluence for different field sizes. It is observed from the results that more the field size, more the dose delivered. However, in case of the 40 × 40 cm^2 field size, the dose obtained was reduced compared with the 30 × 30 cm^2 field size because of the effective area offered by the conical beam, which is less than the square field size. It is also observed that the neutron fluence increases with field size. The neutron fluence spectrum calculated for different field sizes using FLUKA is shown in Fig. 1.17.

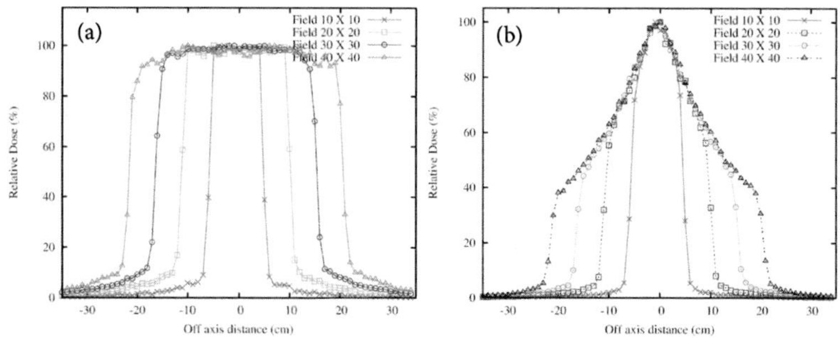

Figure 1.15 (a) Flattened and (b) unflattened dose profile in water phantom for different field sizes.

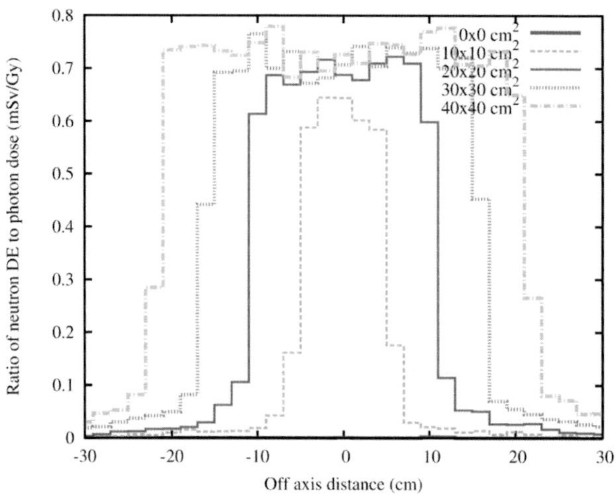

Figure 1.16 The ratio of neutron dose equivalent to central axis photon absorbed dose at patient plane for different field sizes.

Table 1.1 Bremsstrahlung fluence, bremsstrahlung dose delivered in water and neuron fluence for the different filed sizes

Field size (cm^2)	Bremsstrahlung fluence (photon·cm^{-2}/e^-)	Bremsstrahlung dose ($Gy \cdot min^{-1} \cdot \mu A^{-1}$)	Neutron fluence (neutron·cm^{-2}/e^-)	Mean energy of neutron (keV)
10 × 10	7.216×10^{-5}	0.293	2.452×10^{-9}	0.404
20 × 20	8.051×10^{-5}	0.323	3.408×10^{-9}	0.886
30 × 30	8.652×10^{-5}	0.342	3.612×10^{-9}	0.511
40 × 40	8.974×10^{-5}	0.335	3.518×10^{-9}	0.163

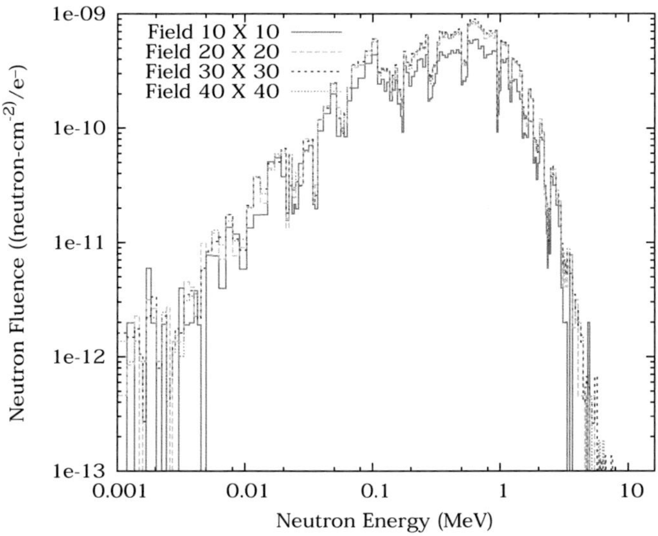

Figure 1.17 The neutron fluence spectra at isocenter for optimized accelerator head assembly.

1.5 Conclusion

The data of bremsstrahlung radiations generated in various targets for different energy of electron from 6 to 18 MeV are summarized and concluded as follows:

- The maximum bremsstrahlung fluence is obtained for a high-Z material (U) and higher incident electron energy (18 MeV).
- The mean energy of bremsstrahlung radiation for a high-Z target is observed more.
- The bremsstrahlung fluence is the maximum in the forward direction and it decreases sharply with angle.
- High-Z targets produce higher neutron fluence for energies greater than 12 MeV electron.
- The contribution of direct transmitted electrons beyond the range of the electron in the target is negligible.

For the production of a 15 MeV photon beam in clinical applications, the design of accelerator head assembly has been proposed and optimized. Using the optimized design, the flattened dose calculated at 100 cm SSD is 0.34 Gy/min at 1 μA for

$30 \times 30\ cm^2$ field size. The maximum square field size that can be produced by the collimator is $30 \times 30\ cm^2$. In addition, the neutron produced in the accelerator head assembly has been estimated and the ratio of the neutron dose equivalent to the gamma dose is found below the allowed limit recommended by the IEC, i.e., <1 mSv/Gy.

To conclude, bremsstrahlung radiations are very useful for medical applications such as radiation therapy and radiography.

Acknowledgment

Dr. Patil is thankful to BCUD, SPPU, Pune, for providing financial assistance under the research project scheme.

References

1. Juliette, T., Jean-Michel, H., Arthur, S. M., Te, V., Jean-Pierre, G. (2013). Past, present, and future of radiotherapy for the benefit of patients. *Nat. Rev. Clin. Oncol.*, **10**, pp. 52–60.
2. Timothy, C. Z., Indra, J. D., Bengt, E. B. (2001). Characteristics of bremsstrahlung in electron beams. *Med. Phys.*, **28**(7), p. 1352.
3. Kenneth, R. H., Peter, R. A. (2006). Review of electron beam therapy physics. *Phys. Med. Biol.*, **51**, pp. R455–R489.
4. Sherman, N. K. et al. (1974). Bremsstrahlung radiators and beam filters for 25-MeV cancer therapy. *Med. Phys.*, **1**(4), p. 185.
5. James, E. R. (2007). Monte Carlo simulations of dose deposition applied to clinical radiation therapy. *Radiat. Meas.*, **41**, pp. S36–S44.
6. Frank, V., Jan, S. (2003). Monte Carlo modelling of external radiotherapy photon beams. *Phys. Med. Biol.*, **48**, pp. R107–R164.
7. Khan, F. M., Gibbons, J. P. (2014). *The Physics of Radiation Therapy*, 5th ed. (Lippincott Williams and Wilkins).
8. Barbalat, O. (1994). *Applications of Particle Accelerators*. CERN documents, CERN-AC-93-04-BLIT-rev.
9. Peter, R. (1975). *Almond Physics Reports* (Section C of Physics Letters) North-Holland Publishing Company, **17**, pp. 11—35.
10. United Nations Scientific Committee on the Effects of Atomic Radiation (UNSCEAR), Sources and effects of ionizing radiation (1993), UN Publications.

11. Alyassin, A. M., Maqsoud, H. A., Mashat, A. M., Al-Mohr, A. S., Abdulwajid, S. (2013). Feasibility study of gamma-ray medical radiography. *Appl Radiat Isot.,* **72**, pp. 16–29.
12. Robin, H., Brendan, H., Lois, H., Zdenka, K., David, T., Clive, B. (2014). Advances in kilovoltage x-ray beam dosimetry. *Phys. Med. Biol.,* **59**(6), p. R183.
13. Podgorsak, E. B. (2005). *Radiation Oncology Physics: A Handbook for Teachers and Students,* IAEA.
14. Mijnheer, B. J., et al. (1987). What degree of accuracy is required and can be achieved in photon and neutron therapy? *Radiother. Oncol.,* **8**, pp. 237–252.
15. Laramore, G. E. (2009). Role of particle radiotherapy in the management of head and neck cancer. *Curr. Opin. Oncol.,* **21**, pp. 224–231.
16. Schulz-Ertner, D., Tsujii, H. (2007). Particle radiation therapy using proton and heavier ion beams. *J. Clin. Oncol.,* **25**, pp. 953–964.
17. Ma, C. M., Maughan, R. L. (2006). Within the next decade conventional cyclotrons for proton radiotherapy will become obsolete and replaced by far less expensive machines using compact laser systems for the acceleration of the protons. *Med Phys.,* **33**, pp. 571–573.
18. Brahme, A. (1988). Accuracy requirements and quality assurance of external beam therapy with photons and electrons. *Acta Oncologica,* Supplementum 1.
19. International Commission on Radiation Measurements and Units (ICRU) (1976). Determination of Absorbed Dose in a Patient Irradiated by Beams of X or Gamma Rays in Radiotherapy Procedures. *ICRU Report 24,* Bethesda, Maryland.
20. Mohan, R., Chui, C., Lidofsky, L. (1985). Energy and angular distributions of photons from medical accelerators. *Med. Phys.,* **12**, pp. 592–597.
21. NCRP (1993). Limitation of exposure to ionization radiation. In *National Council on Radiation Protection and Measurement* (Report No. 116, Washington, DC).
22. Singru, R. M., *Introduction to Experimental Nuclear Physics.* Wiley Eastern Pvt. Ltd.
23. Davisson, C. M. (1965). Interaction of gamma radiation with matter, in Alpha, Beta and Gamma Ray Spectroscopy, K. Siegbahn (Editor), North Holland, Amsterdam, pp. 37–78.

24. Loi, G. et al. (2006). Neutron production from a mobile linear accelerator operating in electron mode for intraoperative radiation therapy. *Phys. Med. Biol.*, **51**, pp. 695–702.
25. Mobley, R. C., Laubenstein, R. A. (1950). Photoneutron thresholds of beryllium and deuterium. *Phys. Rev.*, **80**(3), p. 309.
26. Rajamanickam, B., Kuo, A. L., Richard, Y., Kheng-Wei, Y. (2012). Cancer and Radiation Therapy: Current Advances and Future Directions. *Int. J. Med. Sci.*, **9**(3), pp. 193–199.
27. Fasso, A., et al. (2005). FLUKA: A multi-particle transport code. CERN-2005-10, INFN/TC-05/11, SLAC-R-773.
28. Seltzer, S. M., Berger, M. J. (1986). *At. Data Nucl. Data Tab.*, **35**, p. 345 .
29. Fasso, A., et al. (1995). FLUKA: Performance and applications in the intermediate energy range. Proc. 1st AEN/NEA Specialists' Meeting on Shielding Aspects of Accelerators, Targets and Irradiation Facilities (SATIF 1), Arlington (Texas), Apr. 28–29 (1994). *OECD Documents*, 1995, pp. 287–304.
30. Price, K. W., et al. (1978). Fast and thermal neutron profiles for 25-Mv x-ray beam. *Med. Phys.*, **5**(4), pp. 285–289.
31. Patil, B. J., et al. (2010). Simulation of e^--n targets by FLUKA and measurement of neutron flux at various angles for accelerator based neutron source. *Ann. Nucl. Energy*, **37**(10), pp. 1369–1377.
32. Patil, B. J., Chavan, S. T., Pethe, S. N., Krishnan, R., Dhole, S. D. (2010). Measurement of angular distribution of neutron flux for the 6 MeV race-track microtron based pulsed neutron source. *Appl. Radiat. Isot.*, **68**(9), pp. 1743–1745.
33. Van, L. K. Mondelaerst, W. (1997). Design of field flattening filters for a high-power Bremsstrahlung converter by full Monte Carlo simulation. *Radiat. Phys. Chem.*, **49**(3), pp. 307–317.
34. Verhaegen, F., Seuntjens J. (2003). Monte Carlo modelling of external radiotherapy photon beams. *Phys. Med. Biol.*, **48**, pp. R107–R164.
35. Golnik, N., et al. (2007). Measurement of the neutron dose near a 15 MV medical linear accelerator. *Radiat. Prot. Dosim.*, **126**(1–4), pp. 619–622.

Chapter 2

The Investigation of Cobalt-60 Tomotherapy

Chandra P. Joshi,[a,b] **Pandit B. Vidyasagar,**[c,d] **and L. John Schreiner**[a,b]

[a]*Cancer Centre of Southeastern Ontario, Canada*
[b]*Departments of Oncology and Physics, Queen's University, 25 King Street West, Kingston, Ontario, Canada K7L 5P9*
[c]*SNT Marathwada University, Nanded, Maharashtra, India*
[d]*Department of Physics, University of Pune, Pune, Maharashtra, India*

chandra.joshi@krcc.on.ca

This chapter is a summary of work investigating the clinical potential of Cobalt-60 (Co-60) units for conformal radiation therapy techniques via tomotherapy, originally presented in the doctoral thesis of one of the authors [1]. The thesis work was performed under the supervision of Profs. Vidyasagar and Schreiner. This work utilized user codes of the EGSnrc Monte Carlo Code system to investigate different Co-60 source designs for tomotherapy applications, particularly to determine potential radiation output enhancement for a fan beam with different source shapes (cylindrical and rectangular). Fan beam characteristics such as penumbra, and intensity modulation for various Co-60 source and hypothetical Co-60 tomotherapy unit designs were

Radiation in Medicine and Biology
Edited by Pandit B. Vidyasagar, Sagar S. Jagtap, and Omprakash Yemul
Copyright © 2017 Pan Stanford Publishing Pte. Ltd.
ISBN 978-981-4745-92-5 (Hardcover), 978-1-315-20656-1 (eBook)
www.panstanford.com

investigated. Monte Carlo simulated dose data were used to calculate tomotherapy plans from various hypothetical and existing units for an example head and neck cancer case. Results demonstrate a potential for significant increase in fan beam output for rectangular source shapes. A comparison of intensity-modulated energy fluence profiles showed that irrespective of the source shape and size, the Co-60 source provides adequate intensity modulation required for tomotherapy applications. A dose–volume analysis of tomotherapy plans shows that Co-60 tomotherapy can provide clinically competitive conformal dose delivery. Since the completion of the thesis work, several developments have happened in Co-60-based radiation therapy; this chapter ends with a brief review of further development in Co-60-based radiation therapy.

2.1 Introduction

In a recent World Health Organization report [2], cancer figures among the leading causes of morbidity and mortality worldwide, with approximately 14 million new cases and 8.2 million cancer-related deaths in 2012. The report notes that the number of new cancer cases worldwide is expected to rise by about 70% over the next two decades. More than 60% of the world's total new annual cases occur in low and middle income countries (LMICs) mainly in Africa, Asia and Central and South America; these cases account for 70% of the world's cancer deaths [2, 3]. Radiation therapy (RT) plays an important role in the treatment of cancer with approximately 50% of cancer patients receiving RT treatment in curative, adjuvant, or palliative setting [4]. However, despite being home to 85% of the world's population, there are less than 35% of the world's radiotherapy facilities in LMICs (around 4400 megavoltage machines), leaving most cancer patients in LMICs without any access to potentially life-saving RT treatment [5].

For many decades, the conventional cobalt 60 (Co-60) radiation therapy units were the main workhorse of radiation therapy departments worldwide, particularly in LMICs. The major research and development efforts in modern radiation therapy have been to better conform dose to the target to be treated while minimizing dose to healthy tissue. This improved dose delivery was achieved by developing conformal radiation therapy techniques such as intensity-modulated radiation therapy (IMRT) and tomotherapy—development that was mainly limited to linear accelerators. There was relatively little development in the Co-60 radiation therapy units for the implementation of modern IMRT and tomotherapy techniques. This background is the main motivation for our work, which was mainly undertaken to explore potential improvements in Co-60 units and their clinical viability for modern tomotherapy techniques. Our view was that if technically feasible and clinically competitive, an improved and redesigned Co-60 unit may provide a reliable, low cost, low maintenance and state-of-the-art IMRT. This may be of particular interest to LMICs where cancer patient population is greater but financial resources are scarce.

The relative lack of development of Co-60 radiation therapy units has been attributed to the perceived disadvantages associated with Co-60 radiation such as relatively lower photon energy, larger beam penumbra due to source size, lower radiation output, and periodic source replacement requirements. A Co-60 beam, although of lower photon energy, offers mono-energetic spectra with two clean and higher peaks (1.17 and 1.33 MeV) compared to the spectra of a 6 MV photon beam [6]. It has long been known that beam energy issues such as percent depth dose (or penetration) become insignificant when one moves to the rotational radiation delivery [7, 8]. Johns and Cunningham [1983] demonstrated that differences in the dose distributions achieved in the bladder treatments with 10 MV x-rays and with the Co-60 γ-rays become minor when rotational approaches are used. Other researchers have also reported that with the advanced multi-beam treatment modalities, advantages of very high-energy treatment decrease and become negligible, particularly in the IMRT [9] and tomotherapy [8] techniques. This argument was further supported by dose plans for a typical prostate cancer treatment with Co-60 and 6 MV photon energies using 18-field conventional three-dimensional conformal radiation therapy (3DCRT) techniques presented in Fig. 2.1. These plans show comparable isodose distributions, particularly in the high dose treatment regions in both the Co-60 and 6 MV plans [1].

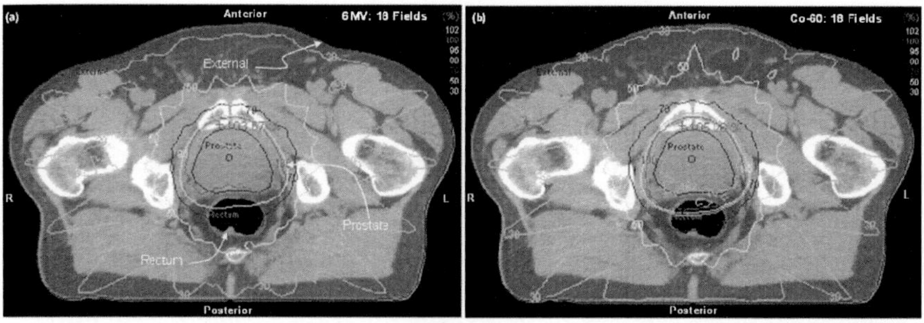

Figure 2.1 Eighteen field 3DCRT treatment plans for cancer of prostate for (a) 6 MV and (b) Co-60 radiations. The plans were calculated on Theraplan Plus treatment planning system (Nucletron, Kanata, ON).

The potential for highly conformal dose delivery has been strongly confirmed recently with first reports from a novel cobalt/MRI unit developed by View Ray (Oakwood, OH, USA, www.viewray.com) [10–12]. In another novel Co-60-related development, the GammaPod™ system (Xcision Medical, Columbia, MD, USA, www.xcision.com) is a dedicated breast cancer-specific stereotactic body radiation therapy unit using multiple Co-60 sources.

In this chapter, we present a brief review of our investigations of efficient Co-60 source designs based on Monte Carlo (MC) simulations. Subsequently, a comparison of MC calculated arbitrary energy fluence profiles was done to evaluate quality of intensity modulations for different hypothetical Co-60 unit models and two existing clinical units (Co-60 and 6 MV). The hypothetical units used rectangular and cylindrical sources with customized binary multi-leaf collimator (MLC) models designed for them, while the clinical units were modeled using standard MIMiC® binary MLCs (Best nomos, Pittsburg, PA). This is followed by a qualitative comparison of tomotherapy plans for a typical head and neck cancer calculated with an in-house dose calculation and optimization program utilizing the MC calculated beamlet data.

2.1.1 Tomotherapy

Tomotherapy is one of the earliest implementations of IMRT [8, 13]. "Tomotherapy" literally means "slice therapy" and employs a radiation source revolving around the patient with a collimated narrow radiation fan-beam that projects a slice on to the patient. The radiation intensity across each slice is modulated using a digitally controlled binary MLC, with leaves closing and opening as the radiation source revolves and the patient is advanced through the ring-shaped gantry. Tomotherapy can be summarized as an "optimized rotational slice intensity-modulated conformal radiation therapy" technique. Tomotherapy dose delivery can be achieved by using two approaches, axial and helical. In axial tomotherapy, one slice is treated per rotation of the intensity-modulated radiation fan beam and the patient is translated in steps. This was implemented, for example, in the first IMRT advancement with linacs using the binary "multileaf intensity-modulation collimator" MIMiC® system [13]. In helical tomotherapy, the radiation source is mounted on a CT-like gantry and the conformal dose delivery is achieved by the continuous translational motion of the patient through the gantry simultaneous with the intensity-modulated radiation fan beam rotations. Currently, the Hi-Art tomotherapy unit (Accuray/TomoTherapy, Madison, WI, USA) using a 6 MV linac x-ray source is the only available helical tomotherapy unit. This unit also has an array of detectors placed opposite to the radiation source in the same ring gantry that can be used for mega-voltage computed tomography (MVCT) imaging. This enables quick imaging useful for treatment planning, routine setup verification and treatment correction/adaptation [14, 15].

2.2 Investigations of an Efficient Co-60 Source Design

The main objective of my thesis work, under the guidance of Professors Vidyasagar and Schreiner, was to investigate different aspects of the development and implementation of cobalt-60-based tomotherapy [1]. One focus was to investigate the potential for increase in fan beam radiation output of Co-60-based tomotherapy unit through change in source design (cylindrical to rectangular), improvement in the packing density of the source and decrease in source to axis (isocenter) distance (SAD).

The EGSnrcMonte Carlo simulations utilizing the BEAMnrc and DOSXYZnrc user codes [16–18] were extensively used to conduct these investigations.

A potential increase in fan beam radiation output for tomotherapy applications could be achieved through an optimal choice and/or combination of the following parameters: (i) the source shape (e.g., cylindrical to rectangular), (ii) the specific activity (Curies/gram) of Co-60, (iii) the packing density of active material in the source volume, (iv) the source volume to increase the source strength, and (v) the source to axis distance (SAD) of the treatment unit.

Many of these modifications have practical and clinical limitations associated with them. For example, the achievable specific activity of an isotope is technically limited by the atomic number of the stable isotope and the activation time in the nuclear reactor. An increase in the packing density, as well as a change in source volume through an increase in the source height, would increase the self-absorption by the source resulting in the loss of radiation output. On the other hand, a larger diameter cylindrical source would lead to an increase of the radiation fluence outside the clinically useful fan beam area. Also, the reduction in the SAD inherent in a tomotherapy unit limits the dimensions of overall source housing (shielding) and the collimation system, and if made too small would limit the physical space requirement for the patient positioning and immobilization. Therefore, not all parameters could be optimized for a potential clinical Co-60 tomotherapy unit.

Non-cylindrical Co-60 source geometries in rectangular (cube and parallelepiped), truncated pyramid and prism shapes were initially considered for the Co-60 source shape studies. It was realized that although a prism-shaped source may be marginally better than a rectangular-shaped (RS) source for fan beam applications due to potentially smaller penumbra, a rectangular source would be technically easier to manufacture. Therefore, our main attention was on rectangular-shaped sources.

Monte Carlo simulations were performed for a conventional Co-60 T780c unit (Best Theratronics, Canada) housing a standard 2 cm diameter cylindrical source and for hypothetical Co-60 units with a 3 cm diameter cylindrical and 3 and 5 cm-long rectangular sources. A brief review of the various source dimensions and used acronyms is given in Table 2.1. The secondary field defining jaws, as used in conventional T780c units, were considered unnecessary for hypothetical Co-60-based tomotherapy units. Thus, a modified collimator system using a thicker primary collimator (10 cm-high lead metal) and an 8 cm-high tungsten binary MLC was modeled in the hypothetical unit simulations. In the modified collimator system, the primary collimator opening was customized for fan beam applications only, and fan beam and beamlet fields were defined by the binary MLC leaves. Using MC simulated dose profiles, fan beam 20–80% penumbra and 50–90% fringe distances were calculated for various combinations of SAD and collimator to isocenter distance. Effects of source shape and increased packing density of active material (including self-absorption by the source) on fan beam radiation output were also investigated.

2.2.1 Description and Validation of the New Source Code

The existing BEAMnrc MC code only allowed for the modeling of isotropic radioactive sources in a point or cylindrical (CYL) source geometries [17]. In order to facilitate the MC modeling of isotropic non-cylindrical Co-60 source designs, a new MC code was written to model the isotropic radiation source in six-faced polyhedron geometries. This new source code was integrated with the BEAMnrc code. The new source code can be used to simulate isotropic radioactive sources in rectangular (e.g., cube, parallelepiped), prism and truncated pyramid geometries. Extensive quality assurance tests were performed to validate the new source code using independent simulations with BEAMnrc and DOSXYZnrc codes [17, 18]. These validations included verifications of isotropic nature, randomness of the particle origins within the source and their travel paths, photon spectra and depth dose calculations, etc., for a rectangular source modeled with the new source code. The validation results were all published [6].

2.2.2 Effect of Rectangular-Shaped Co-60 Source Width on Fan Beam Output

The potential of fan beam output enhancements with the different cylindrical and rectangular source shapes was investigated using MC simulations. The 20 × 1 cm^2 fan beam outputs were calculated for different dimensions of rectangular sources and compared with the fan beam outputs for the 2 and 3 cm diameter cylindrical sources (see Table 2.1 for all source dimensions and acronyms). In all simulations, a standard packing density of 5.88 g/cm^3 was assumed, which is equal to the effective packing density of the conventional cylindrical source used in a T780c unit. Absolute dose profiles at 5 cm depth in water for 20 × 1 cm^2 fan beams were extracted from BEAMnrc/DOSXYZnrc simulations [18]. The dose profile data was normalized to 100% at the beam central axis from the 2 cm diameter cylindrical source of the conventional T780c Co-60 unit. The normalized dose profiles presented in Fig. 2.2 demonstrate that for a fixed source volume (8.8 cm^3) a significant increase in fan beam radiation output can be achieved merely by a change in the source shape.

Table 2.1 Acronyms and dimensions of geometrical shapes of investigated Co-60 sources.

	Rectangular shaped (RS) sources		**Cylindrical (CYL) sources**		
Acronym →	**RS_L3 cm**	**RS_L5 cm**		**CYL_D2 cm**	**CYL_D3 cm**
Length (*Along the fan beam*)	3 cm	5 cm	Diameter	2 cm	3 cm
Height (*Normal to the fan beam*)	2.09 cm	1.26 cm	Height (*Normal to the fan beam*)	2.8 cm	1.25 cm
Thickness (*Across the fan beam*)	1.4 cm	1.4 cm			

The data presented in Fig. 2.3 show that fan beam output enhancements of up to 19% and 30%, compared to a 2 cm diameter cylindrical source, can be achieved by using 3 and 5 cmlong rectangular-shaped sources, respectively. The influence of increasing Co-60 packing density was also investigated. Figure 2.3 shows the overall projected gains in fan beam output for different shapes and sources for different packing densities

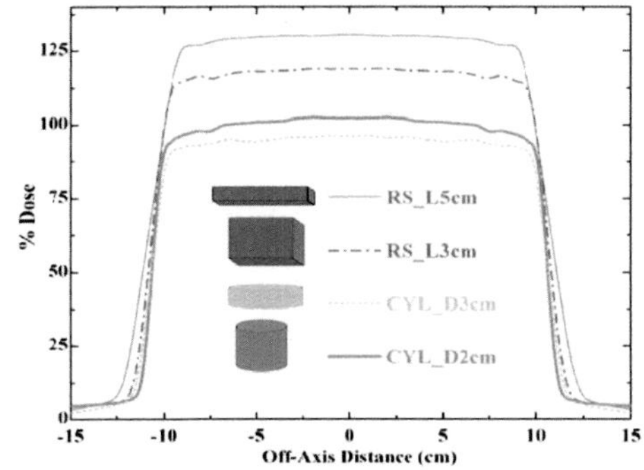

Figure 2.2 Simulated fan beam (20 × 1 cm^2) dose profiles (along the long axis) at a depth of 5 cm in water (collimator to isocenter distance 23.5 cm, SAD of 80 cm) for different Co-60 sources of cylindrical (CYL) and rectangular-shaped (RS) geometries. The dose profiles are normalized to the dose profile of a 2 cm diameter cylindrical source (CYL_D2 cm) used in a conventional T780c unit. The inset shows the proportionate diagrams of the sources in relation to one another.

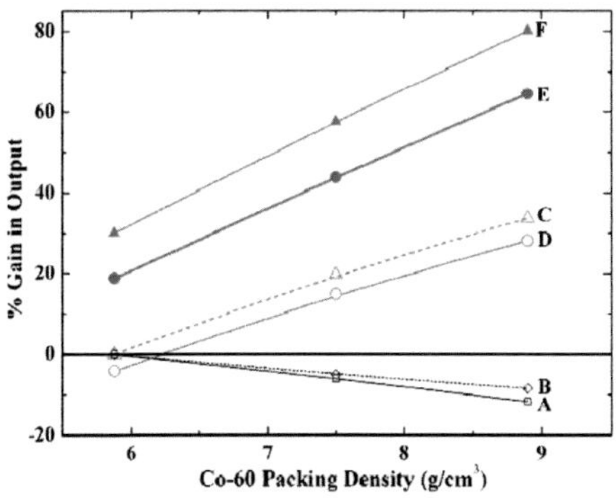

Figure 2.3 Estimate of potential increases in fan beam output for Co-60 source packing densities between 5.88 and 8.90 g/cm^3. All the sources have active volumes of 8.8 cm^3. *Self-Absorptions*: Lines A and B represent the percentage self-absorption for a 2 cm diameter cylindrical source (CYL_D2 cm) and a 3 cm-long rectangular-shaped source (RS_L3 cm), respectively. *Output Gains*: Lines C, D, E and F represent potential percentage gains in outputs for 2 and 3 cm diameter cylindrical sources (CYL_D2 cm and CYL_D3 cm) and 3 and 5 cm-long rectangular sources (RS_L3 cm and RS_5 cm). All fan beam outputs are normalized to the fan beam output at SAD of 80 cm of a conventional T780c unit using a 2 cm diameter cylindrical source with 5.88 g/cm^3 packing density.

of Co-60 sources. These gain estimates do account for self-absorption and the change in source shape. All gain estimates are normalized to the fan beam output of a conventional Co-60 unit with a 2 cm diameter cylindrical source. The data presented in Fig. 2.3 demonstrates that at an intermediate packing density of 7.5 gm/cm^3, overall gains of ~20% and ~15% are possible for the 2 and 3 cm diameter cylindrical sources, respectively. With 3 and 5 cm-long rectangular sources, overall gains increase to 44% and 58%, respectively, with the same packing density (Fig. 2.3). These results emphasize the potential for improvement in source design to enhance the output of a Co-60 unit for fan beam applications.

2.2.3 Penumbra and Fringe Distance Estimates for Different Hypothetical Units

For this particular aspect of the study, hypothetical Co-60 units with 3 and 5 cm-long rectangular sources with different combinations of collimator to isocenter distances (17, 25, 30, 35, and 40 cm) and SADs (60, 70 and 80 cm) were modeled [6]. These unit models were schematically similar to the models shown in Figs. 2.4a,b. For these units, 20 × 1 cm^2 fan beam dose profiles at a depth of 10 cm in water were calculated using MC simulations. Estimates of fan beam (20–80%) penumbra and the (50–90%) fringe distances for various hypothetical Co-60 tomotherapy units utilizing different combinations of rectangular source size, collimator to isocenter distances and SADs were obtained (Fig. 2.5). Overall the data presented in Fig. 2.5 provide a useful tool for the selection of the optimal combination of source size, source shape, and treatment unit collimator to isocenter distance and SAD for the design of a Co-60-based tomotherapy unit. This modeling shows that units with large source size, larger collimator to isocenter distances and shorter SADs, give larger beam penumbra than appropriate for clinical applications. Also very short collimator to isocenter distances and SAD would hinder patient positioning as distances between the treatment couch and the source head become restrictive. The dotted box area in Fig. 2.5 shows the region of practical clinical design. Thus a relatively generous collimator to isocenter distance of 30 cm was used in all subsequent modeling.

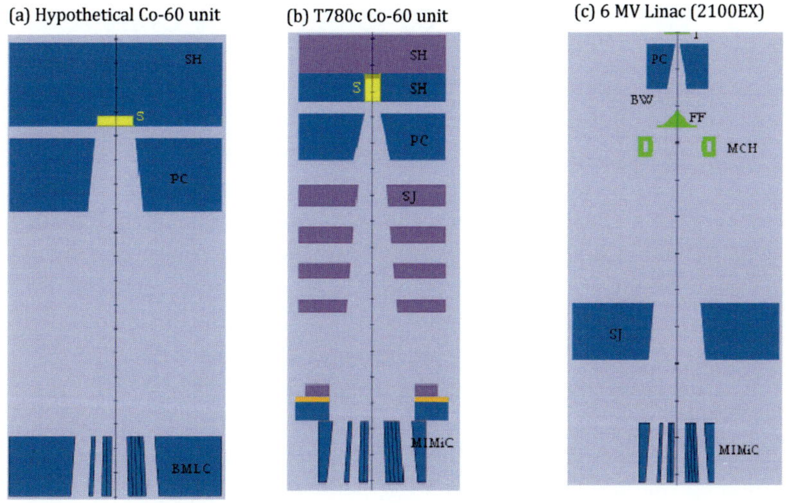

Figure 2.4 The geometry of the various units modeled in these Monte Carlo studies. (a) A hypothetical Co-60 unit with a custom binary MLC. (b) A T780c unit with the standard MIMiC®. (c) A 6 MV linac with the standard MIMiC®. S = source, SH = shielding, PC = primary collimator, SJ = secondary jaws, T = target, FF = flattening filter, MCH = monitor chamber, BW = beryllium (vacuum) window.

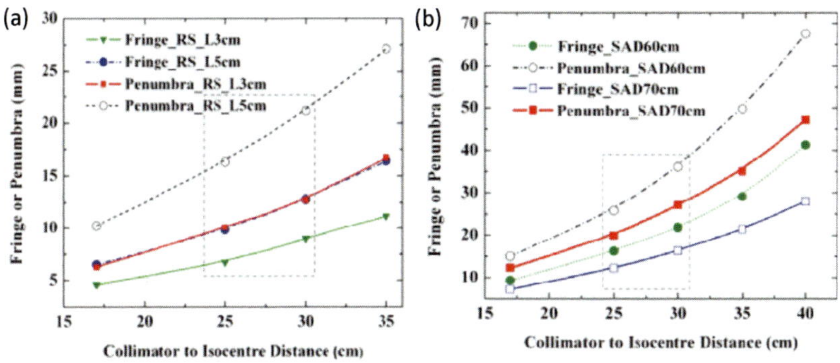

Figure 2.5 The fan beam fringe (50–90%) and penumbra (20–80%) distances for different collimator to isocenter distances. The boxed areas represent technically practical and clinically acceptable penumbra and collimator to isocenter distances: (a) for 3 cm (RS_L3 cm) and 5 cm (RS_L5 cm)-long rectangular-shaped sources at the SAD of 80 cm and (b) for 5 cm-long RS source (RS_L5 cm) at the SADs of 60 and 70 cm.

2.3 Hypothetical Co-60 Tomotherapy Units

Further investigations were performed to study the intensity modulation that could be achieved by various Co-60 unit designs. These designs included five hypothetical Co-60 tomotherapy units with rectangular and cylindrical-shaped Co-60 sources along with a clinical T780c Co-60 unit (Best Theratronics, Kanata, ON, Canada). A CL2100Ex 6 MV x-ray linear accelerator (linac) (Varian Medical Systems, Palo Alto, CA, USA) was included for comparison. The energy fluence profiles and intensity-modulated fan beams were then extracted directly from the BEAMnrc [17] MC calculated data. Where required, the DOSXYZnrc MC User code [18] was used to model the dose in water from the phase-space data extracted from the BEAMnrc MC simulations. In order to keep the statistical errors ≈1%, sufficient histories (2×10^9 to 20×10^9 in BEAMnrc and 0.05×10^9 to 1×10^9 in DOSXYZnrc) were used in the simulations.

The hypothetical units with 3 and 5 cm-long rectangular-shaped Co-60 sources were modeled at SADs of 70 and 80 cm (Fig. 2.4a) for this part of the investigation. A hypothetical unit with a 3 cm diameter cylindrical Co-60 source was also modeled at SAD of 80 cm. In a dedicated Co-60 tomotherapy unit, the radiation field in the treatment plane would likely be defined by the binary MLC leaves, as discussed before the secondary field defining jaws were considered unnecessary. All hypothetical units were modeled using individual custom binary MLC designs with collimator to isocenter distances set to 30 cm (Fig. 2.4a). The conventional T780c Co-60 unit (Fig. 2.4b) and a 6 MV linac (Fig. 2.4c) were modeled with the standard MIMiC® binary MLCs. The collimator to isocenter distances for the T780c Co-60 unit and the 6 MV linac were 30 and 39 cm, respectively.

2.3.1 Intensity-Modulated Fan Beam Energy Fluence Profiles

In IMRT, fluence profiles with sharp variations in energy fluence (i.e., with more prominent peaks and valleys) indicate a better ability to modulate the intensity within a beam. To a large extent, the sharp variations within a beam are dependent on the size of the radiation source as well as the binary MLC design. A treatment unit with the ability to produce sharper variation within an intensity-modulated beam would be expected, a priori, to provide a higher degree of conformality in treatment planning. Therefore, for evaluation purposes, the energy fluence profiles with sharper variations (i.e., modulation) within the fan beam are considered to be qualitatively superior.

The potential for clinical fan beam intensity modulations was studied for these seven different tomotherapy units including five hypothetical Co-60 units and a conventional Co-60 and a 6 MV linac. The MC calculated intensity-modulated energy fluence profiles utilizing the same set of arbitrary fan beam segments were extracted from the BEAMnrc data for these seven units. Figures 2.6a–f present a comparison of intensity-modulated relative energy fluence profiles from the 6 MV linac (CL2100EX), 2 cm diameter cylindrical source (T780c), hypothetical units with 3 cm cylindrical, and 3 and 5 cm-long rectangular Co-60 sources. These profiles are taken in air in the plane of the isocenter of the unit. Co-60 units had SADs of 70 and 80 cm, while the 6 MV linac had an SAD of 100 cm. The relative energy fluence profiles are normalized to 1.0 at the central axis of the largest possible fan beam segment (20×1 cm^2).

The qualitative comparison of energy fluence profiles (Figs. 2.6a–d) demonstrates that Co-60 units with 2 cm diameter cylindrical and 3 cm-long rectangular-shaped Co-60 source provide superior intensity modulation than the 5 cm-long rectangular Co-60 source. This can be attributed to significantly larger Co-60 source size (5 cm) compared to the size of an individual beamlet (1×2 cm^2). Figures 2.6a,b show that the 6 MV x-ray source (on a linear accelerator) offers a superior intensity modulation than that by Co-60 sources. Figures 2.6e,f review the intensity-modulated relative energy fluence profiles at SADs of 70 and 80 cm for hypothetical Co-60 units with 3 and 5 cm-long rectangular-shaped sources. A qualitative comparison of these profiles shows a small (but not negligible) loss in the quality of intensity modulation at shorter SADs. This can be observed from relatively larger penumbras and shallower valleys in fluence profiles for the units with a shorter SAD (70 cm). Also, intensity modulation with the 3 cm-long rectangular source is qualitatively superior compared to that from the 5 cm-long rectangular source (Fig. 2.6c). Overall, the results show that the ability of the intensity modulation to achieve desired conformality of dose delivery deteriorates with increase in the size of the source (Figs. 2.6a–f) as well as at shorter SADs (Figs. 2.6e,f). However, optimal source geometry for tomotherapy applications will depend on a careful evaluation of the resulting tomotherapy plans from units with these sources. This was the next step in our studies.

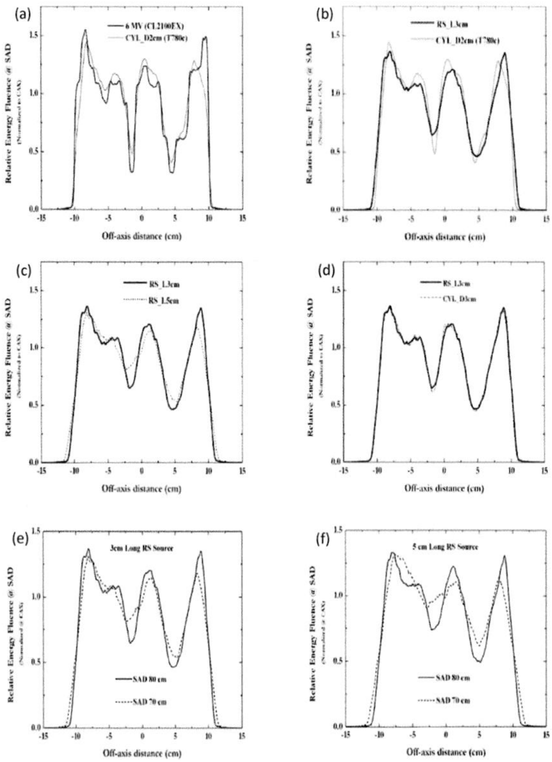

Figure 2.6 Comparison of intensity-modulated relative energy fluence profiles obtained by summing the energy fluence profiles for individual segments produced with binary MLCs from: (a) A 6 MV linac and T780c Co-60 unit. (b) A T780c unit and the hypothetical unit with the 3 cm-long rectangular-shaped source; 6 MV linac has an SAD of 100 cm, Co-60 unit SADs were set to 80 cm. (c) Hypothetical units with 3 and 5 cm-long rectangular-shaped (RS) sources (d) Hypothetical units with 3 cm-long rectangular and 3 cm diameter cylindrical source; in both of the Co 60 units, the SAD were set to 80 cm. (e) 3 cm-long RS, SAD, 70 and (f) 5 cm-long rectangular sources for SADs 70 and 80 cm.

2.4 Tomotherapy Treatment Planning

To establish the clinical impact of the quality of intensity modulations offered by each of the seven different units, tomotherapy dose distributions were calculated for a typical head and neck cancer case and analyzed. These dose distributions were compared to evaluate the overall effect of differences in quality of intensity modulation on optimized tomotherapy treatment plans.

The choice of head and neck (H&N) treatment for our studies was made because conventional 3DCRT techniques are often very challenging at this anatomical site. These challenges arise because of the location of various malignant target structures (e.g., gross tumor volume (GTV), clinical target volume (CTV) and involved lymph nodes, etc.) close to the critical normal structures (e.g., the eyes, spinal cord, optic chiasm, parotid, etc.) (Fig. 2.7). Up to the late 1990s, H&N cases treated with conventional 3DCRT techniques often required parallel opposed lateral (bilateral) photon fields to be combined with reduced and adjacent anterior photon fields, and posterior neck electron fields, in the second phase of the treatment, to treat the posterior neck lymph nodes to an adequate dose yet limit the dose to the spinal cord. The field junctions between anterior photon and posterior neck electron fields in these plans presented tremendous problems for a homogeneous dose delivery in the target regions because of the large electron field penumbra. Also, because of dose limitations to the spinal cord, lower energy electron beams were often used resulting in potential under dosing of the posterior neck lymph nodes. The IMRT and tomotherapy techniques presented a tremendous opportunity to provide highly conformal as well as seamless dose delivery in the treatment of H&N cancers.

Tomotherapy dose distributions on a single H&N computed tomography (CT) scan were generated with our five hypothetical Co-60 tomotherapy units, each equipped with a different custom binary MLC system with design dimensions as modeled earlier. Figure 2.7 illustrates the anatomical relationships and the nomenclature used to describe typical targets and avoidance structures on a single head and neck CT scan used in this study. This scan was obtained from a series of typical patient treatment planning CT scans from our clinical practice. For the tomotherapy planning, the same dose–volume optimization and inverse planning objectives were used for all seven tomotherapy plans considered in this study. The CTV represents the gross primary cancer and its microscopic spread. In this particular case, the right posterior neck lymph node (Right Node) was considered to be positive for malignant disease; while the left posterior neck lymph node (Left Node) was considered as a high risk volume for the potential spread of the cancer. A dose of 70 Gy to the CTV was prescribed to treat the primary cancer while doses of 66 Gy to the right node and 50 Gy to the left node were considered adequate for treatment of the nodal disease. A variation of ±5% between the prescribed and delivered doses in the CTV and the nodal regions was considered acceptable. The spinal cord (cord) is a critical organ at risk (OAR), for tomotherapy optimization the maximum dose limit to the spinal cord set to 40 Gy.

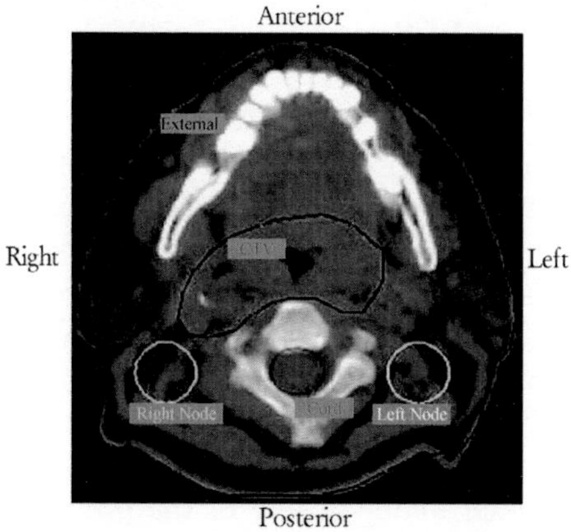

Figure 2.7 A typical H&N case used for tomotherapy planning. CTV = clinical target volume (70 Gy ± 5%), Right Node = right posterior neck lymph node (66 Gy ± 5%), Left Node = left posterior neck lymph node (50 Gy ± 5%) and the spinal cord (≤40 Gy).

2.4.1 Dose Calculation Program and Monte Carlo Simulations

After this initial work an in-house inverse planning and dose–volume optimization program [19] was used to generate the tomotherapy dose distributions. This program is based on an aggressive active set conjugate gradient, i.e., (AAS)CG, algorithm for dose–volume optimization purposes [20]. The program [19] directly utilized the MC calculated dose profile data for individual beamlets emanating from the openings of individual leaf-pairs of the binary MLC or from the standard MIMiC® system. Therefore, in the treatment plans presented here, the MC calculated true individual beamlet dose profiles on the specific head and neck CT scan (Fig. 2.7) for each treatment unit were used for the dose–volume optimization and inverse planning purposes. Note at this stage we did not account for tissue inhomogeneity corrections in the dose calculations. For each of the seven tomotherapy units, 51 beam directions and 16 individual beamlets per beam direction were used for the inverse planning and dose–volume optimization. A matrix size of 2×2 mm^2 was used for the dose calculations. These Co-60 and 6 MV tomotherapy plans were analyzed and compared.

Tomotherapy dose plans for different hypothetical Co-60 units with custom binary MLCs, conventional T780c and 6 MV linac units with a MIMiC® binary MLC were analyzed by comparing isodose distributions, DVHs and dose statistics (Figs. 2.8 and 2.9). Figures 2.8a,b show the tomotherapy dose distribution from the Co-60 unit with the 2 cm diameter source (T780c) and the 3 cm diameter (hypothetical unit) cylindrical sources positioned at SADs of 80 cm. Figures 2.8c–f present the tomotherapy isodose distributions for hypothetical Co-60 units with 3 and 5 cm-long rectangular sources at two different SADs of 70 and 80 cm. A visual and qualitative inspection of dose distributions demonstrate highly conformal high dose regions covering the target volumes (CTV, right and left nodes), while low dose regions surround the spinal cord region. A comparison of dose statistics (*not shown here*) of all these plans show that each of these plans meets the dose criteria set for the target (CTV and nodes) as well as the OAR structures (i.e., cord and normal tissue). In all tomotherapy plans, the maximum doses in the normal tissue region (represented by external body contour minus all other contours) varied between 69.3–70 Gy, hile the average doses in the same region varied between 25.9–35.5 Gy. The dose distributions (Figs. 2.8 and 2.9) demonstrate that the average dose in the normal tissue region is higher for the units with larger sources and shorter SADs.

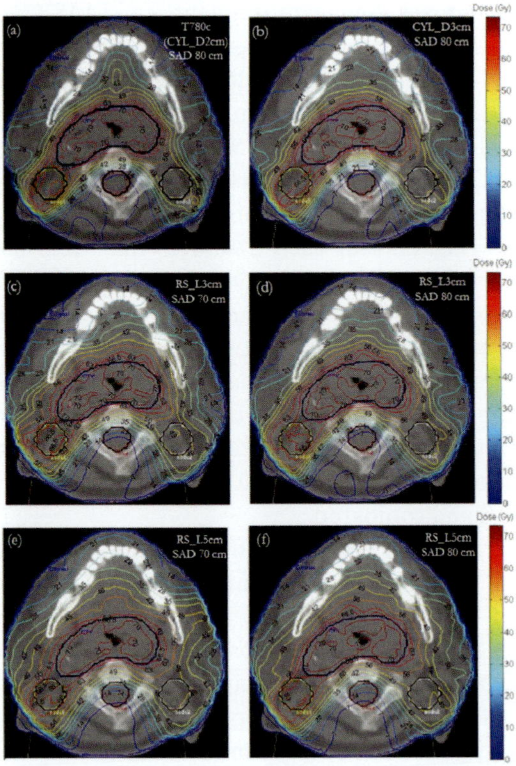

Figure 2.8 Tomotherapy dose distributions in Gy for various Co-60 unit designs (a) the T780c unit with a 2 cm diameter cylindrical source, SAD 80 cm using the standard MIMiC® system; and for the hypothetical units using custom binary MLCs for (b) 3 cm diameter cylindrical source, SAD 80 cm; (c) 3 cm-long rectangular (RS) source, SAD 70 cm; (d) 3 cm-long rectangular source, SAD 80 cm; (e) 5 cm-long rectangular source, SAD 70 cm; and (f) 5 cm-long rectangular source, SAD 80 cm.

Figure 2.9 presents the dose distributions for the tomotherapy plans from the 6 MV and T780c units with MIMiC® binary MLCs. Both units are able to deliver highly conformal treatment doses around the target structures and low doses in the OAR regions. An analysis of the dose statistics show that both plans deliver similar doses to the CTV, nodes, spinal cord and other normal tissue regions. These plans were further compared using DVHs (Fig. 2.9c) and the dose difference map (Fig. 2.9) showing similar regions of high and low doses between the plans. Results shown in Fig. 2.9 demonstrate that an existing clinical Co-60 unit e.g., T780c may require minimal modification such as retrofit of a MIMiC® collimator to produce a clinically acceptable and comparable tomotherapy plan to a MIMiC® equipped 6 MV linac.

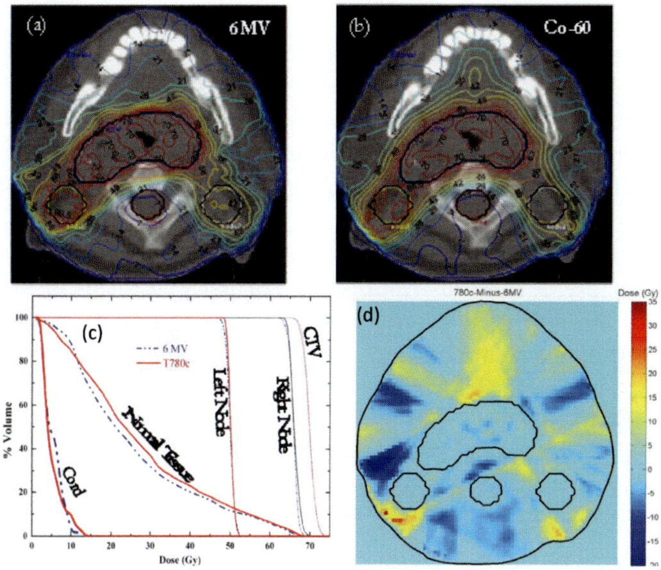

Figure 2.9 A review of the optimized tomotherapy plans for (a) 6 MV linac and (b) the conventional T780c Co-60 unit incorporating the MIMiC® binary MLC. (c) DVH for the tomotherapy plans from 6 MV linac and T780c Co-60 units with MIMiC® MLCs. (d) The dose difference map: T780c plan minus 6 MV plan. Please note that in the dose difference map the yellow and green regions illustrated higher doses (in Gy) with the T780c plan, the blue regions represent the higher doses with the 6 MV plan and the regions with background color represent the region with negligible dose difference between the two plans.

2.5 Discussion

The tomotherapy plans for Co-60 units using sources with similar dimensions but different shapes were compared to investigate the dose distribution differences due solely to the shape. The analyses demonstrate that cylindrical and rectangular-shaped sources with the same maximum dimensions along the fan beam (i.e., 3 cm diameter and 3 cm length) provide qualitatively similar intensity modulation and dose–volume optimization (Figs. 2.6d and 2.9c,d). However, as previously demonstrated in Figs. 2.5 and 2.6, the rectangular source is more efficient than the cylindrical-shaped sources in terms of fan beam radiation output. Therefore, the delivery from the unit with the rectangular source would take less time to achieve the desired tomotherapy treatment.

The overall dose–volume analysis of treatment plans for the seven different tomotherapy units (one 6 MV linac and six Co-60 units) demonstrate that all plans meet the dose objectives originally set for the targets (CTV, right and left nodes) as well as the avoidance structures (spinal cord and normal tissue). Although all these plans are clinically acceptable, the quality of conformality of dose distribution (particularly in terms of overall dose to the normal tissue regions), varies with size of the source and SAD of the treatment unit. Considering the gains in fan beam output with rectangular-shaped sources compared to the output from the T780c unit with the 2 cm diameter cylindrical source, the overall treatment times required for the tomotherapy plans with the rectangular sources will be significantly lower compared to that with the T780c unit. The hypothetical units with 3 and 5 cm-long rectangular sources at SADs of 80 cm are likely to take 19% and 30% lesser treatment times than the T780c unit, respectively. On the other hand, a similar unit with the 3 cm diameter cylindrical source, at 80 cm SAD, will take about 4% more treatment time than the T780c unit. As noted previously, the treatment times will be reduced by an additional 30.6% for the hypothetical units at the SAD of 70 cm due to inverse square effect.

An analysis of the various Co-60 tomotherapy dose plans indicate that despite relative gains in fan beam outputs, a 9% gain with the increase in length of the rectangular source from 3 to 5 cm and an additional gain of 30.6% on the decrease in SAD from 80 to 70 cm, compromises in the conformality of the dose distributions may limit the practical adoption of these strategies. For example, the dose plan with the 5 cm-long source has poorer dose conformality than achieved by irradiation from the 3 cm-long rectangular source, particularly due to the doses delivered in the normal tissue region. Therefore, a change towards a 5 cm-long rectangular source is not well indicated.

The tomotherapy dose distributions for the units presented here are based on the individual beamlets emanating through individual leaf-pairs of the binary MLC [19]. This limitation results from the in-house developed dose–volume optimization program utilizing individual beamlet dose data to generate the optimized dose distribution. Therefore, the distributions presented and discussed here are most likely diminished by exaggerated penumbra effects. Further improvement in the tomotherapy dose distributions is possible with calculations using a more advanced 3D tomotherapy treatment planning system that is able to accurately characterize and model the large Co-60 sources and the binary MLCs. This may mean that dose distributions with large sources could be more conformal than the ones presented here. However, the results presented here are a good indication of the features of Co-60 unit design that affect conformality of the dose delivery.

2.6 A Brief Review of Recent Developments in Co-60-Based RT

Despite the promise and potential of Co-60-based tomotherapy, research in this area thus far has not translated into development of a commercial clinical Co-60 tomotherapy unit. Even so, Co-60 radiation therapy has advanced the course. A brief discussion of the recent progress is presented below.

Dhanesar et al. [21] have developed an aperture superposition dose model algorithm to accurately calculate dose distributions for finite sized Co-60 sources. This is a significant advance improvement for modeling of Co-60 sources in radiation treatment planning systems. The majority of planning system source modeling algorithms mainly cater to linac-based x-ray sources, Co-60 source modeling on these systems has been challenging. Dhanesar et al. have used the aperture superposition dose model for Co-60 extend to 3D tomotherapy dose calculations [21–23].

Recently, significant breakthroughs have happened resulting in technological and commercial development of highly specialized clinical Co-60-based radiation therapy units such as the Tri-Co-60/MRIViewRay System (ViewRay, Inc., Oakwood, OH, USA, www.viewray.com) [10–12] and the GammaPod™ system by the

Xcision Medical Systems (Columbia, MD, USA, www.xcision.com) (see reference [24]).

The ViewRay system (VRS) uses a rotating gantry assembly with three separate Co-60 source heads, three MLCs and a 0.35 Tesla MR imager. The VRS integrates fulltime MR image guidance, Co-60-based IMRT delivery, and intelligent software automation. The system has a tremendous ability to visualize soft tissue on MRI to deliver real time image guided gated radiotherapy on a live anatomy during the radiation delivery [www.viewray.com]. In another novel development, the GammaPod™ system (Xcision Medical) is a dedicated breast cancer-specific stereotactic body radiation therapy unit using multiple Co-60 sources. The Gamma Pod™ system comprises a hemispherical source carrier containing 36 Co-60 sources; the source assembly rotates in synchrony with the collimator system which offers apertures of 1.5 or 2.5 cm, producing 36 non-coplanar, concentric arcs focused at the isocenter. The patient is treated in prone position with the breast positioned inside the hemispherical assembly and immobilized using negative pressure [www.xcision.com].

Also, in the area of conventional Co-60-based radiation therapy, there have been important advancements in conventional Co-60 units such as availability of MLCs, improved patient support systems and computerized control consoles and asymmetric jaws on Co-60 units. Best Theratronics (Kanata, Canada, www.theratronics.ca), traditionally a major manufacturer of Co-60 teletherapy units has heavily invested in the Theratron® Equinox™ unit, the most advanced member of the Theratron® line of Co-60 units. The Equinox unit is available with Multileaf Collimator 60 (MLC 60) option, which is capable of 3DCRT and IMRT deliveries. A commercially enhanced version of Bhabhatron, originally developed by the Bhabha Atomic Research Centre (BARC, Mumbai, India), the Bhabhatron-II Co-60 unit (Panacea Medical Technologies, Bangalore; www.panaceamedical.com) is in clinical use at many cancer clinics in India and other LMICs. The Co-60 units, Equinox (Best Thertronics), and Bhabhatron-II and G-RAY (Panacea Medical) are available with several clinically important enhancements thus far usually available with linear accelerators, including features such as motorized wedges, auto-setup routines, Dicom-RT connectivity, anti-collision mechanisms, and remote diagnostic facilities; development work on integrating electronic portal imaging devices is also underway. Cobalt-60 units are also well suited for total body irradiation (TBI), a low dose rate RT, primarily used in conditioning regimens before bone marrow transplantation. Recently available, GammaBeam^{TM}500 Total Body Irradiator (Best Theratronics, Kanata, Canada) provides a simple, reliable and pragmatic solution to deliver TBI at a large uniform rectangular radiation field at extended source to skin distance of 220 cm.

Cobalt-60-based radiation therapy units have tremendous potential in the treatment of cancer in both conventional and state-of-the-art settings. These developments have shown that if new ideas and novel approaches are instilled, Co-60-based technology can be harnessed to produce unique and potent RT units. We should recognize that radiation therapy facilities in different parts of the world, depending on their demographic location, resources, and clinical needs, may require different levels of sophistication in Co-60 units. Considering limited access to RT, many clinics in certain regions of LMICs may be well served with relatively simpler, practical and less expensive Co-60 units. The clinical use of highly specialized and sophisticated but relatively expensive, but affordable, Co-60 units may meet clinical requirements of RT clinics in developed countries [2–5, 25]. A resurgence of the clinical use of Co-60 RT units in developed countries will help in dispelling misperceptions about Co-60-based RT, potentially encourage embrace of Co-60-based RT in LMICs, particularly in places where access to clinically appropriate and affordable RT is limited.

2.7 Conclusions

Overall, investigations in this work demonstrate that a Co-60 tomotherapy unit could provide adequate intensity modulation required for tomotherapy applications. The development of a Co-60-based tomotherapy unit can offer low cost, less complex and dosimetrically acceptable IMRT solutions via tomotherapy. Therefore, if technically feasible, an improved and redesigned Co-60 unit may provide reliable, low cost, low maintenance and state-of-the-art IMRT via tomotherapy. This may be of particular interest to the LMICs where cancer patient population is greater, availability of radiation therapy facilities is low, and the financial resources are scarce. Also a review of recent advances shows that Co-60-based radiation therapy units have tremendous potential in the treatment of cancer in both conventional and state-of-the-art settings.

Acknowledgments

This chapter is mainly based on excerpts from the PhD thesis titled "The investigation of theoretical, practical and clinical considerations in Cobalt-60 tomotherapy," submitted to the University of Pune [1]. The authors also wish to thank Dr. Johnson Darko and Dr. Sandeep Dhanesar for their collaboration during the thesis work and in our continuing research.

References

1. Joshi, Chandra P. The investigation of theoretical, practical and clinical considerations in cobalt-60 tomotherapy PhD Thesis: Supervisors—Prof. P. B. Vidyasagar and Prof. L John Schreiner (University of Pune, Pune , 2008).
2. Stewart, B. W., and C. P. Wild. World Cancer Report 2014. Lyon, France: International Agency for Research on Cancer. World Health Organization (2014).
3. World Health Organization. (2012). GLOBOCAN 2012: Estimated Cancer Incidence. Mortality and Prevalence Worldwide in 2012.
4. Barton, M. B., Michael F., and Jesmin, S. (2006). Role of radiotherapy in cancer control in low-income and middle-income countries. *The Lancet Oncology*, **7**, p. 584.
5. Samiei, M. (2013). Challenges of making radiotherapy accessible in developing countries. *Cancer Control*, **85** (http://cancercontrol.info/wp-content/uploads/2014/08/cc2013_83-96-Samiei-varian-tpage-incld-T-page_2012.pdf).
6. Joshi, C. P., Darko, J., Vidyasagar, P. B., and Schreinerm, L. J. (2008). Investigation of an efficient source design for cobalt-60-based tomotherapy using EGSnrc Monte Carlo simulations. *Phys. Med. Biol.*, **53**, p. 575.
7. Johns, H. E., and Cunningham, J. R. (1983). *The Physics of Radiology* (Springfield, IL: Charles C Thomas).

8. Mackie, T. R., Holmes, T. W., Swerdloff, S., Reckwerdt, P., Deasy, J. O., Yang, J., Paliwal, B., and Kinsella, T. (1993). Tomotherapy: A new concept for the delivery of conformal radiotherapy. *Med. Phys.*, **20,** p. 1709.
9. Sternick, E. S., et al. (1997). Intensity modulated radiation therapy: What photon energy is best. Presentation at the XIIth International Conference on the Use of Computers in Radiation Therapy (ICCR), Leavitt, D. D., et al., ed. (Madison, WI: Medical Physics Publishing), pp. 418–419.
10. Dempsey, J. F. (2011). System for delivering conformal radiation therapy while simultaneously imaging soft tissue. U.S. Patent No. 7,907,987.
11. Mutic, S., Dempsey, J. (2014). The view ray system: Magnetic resonance–guided and controlled radiotherapy. *Semin. Radiat. Oncol.*, **24** pp. 196–199.
12. Wooten, H. O., Rodriguez, V., Green, O., Kashani, R., Santanam, L., Tanderup, K., Mutic, S., and Li, H. (2015). Benchmark IMRT evaluation of a Co-60 MRI-guided radiation therapy system. *Radiotherapy and Oncology*, **11,** p. 402.
13. Carol, M. P. (1995). Peacock: A system for planning and rotational delivery of intensity modulated fields. *Int. J. Image Syst. Technol.*, **6,** p. 56.
14. Mackie, T. R., Kapatoes, J., Ruchala, K., Lu, W., Wu, C., Olivera, G., et al. (2003). Image guidance for precise conformal radiotherapy. *Int. J. Radiat. Oncol. Biol. Phys.*, **56,** p. 105.
15. Yartsev, S., Kron, T., and Van Dyk, J. (2007). Tomotherapy as a tool in image-guided radiation therapy (IGRT): Theoretical and technological aspects. *Biomed. Imaging Interv. J.*, **3**(1), p. e15.
16. Kawrakow, I., and Rogers, D. (2000). The EGSnrc code system: Monte Carlo simulation of electron and photon transport. Technical Report PIRS–701 (Ottawa, Ontario: National Research Council of Canada).
17. Rogers, D. W. O., Walters, B., and Kawrakow, I. (2005). BEAMnrc Users Manual, *NRCC Report PIRS-0509 (A) rev I* (Ottawa, ON: IRS, NRCC).
18. Walters, B., Kawrakow, I., and Rogers, D. W. O. (2005). DOSXYZnrc users manual. *NRC Report PIRS*, p. 794.
19. Dhanesar, S. C., Rogers, M., Joshi, C., Darko, J., Salomons, G., Kerr, A., and Schreiner, L. J. (2007). Experimental validation of conjugant gradient

based inverse treatment planning system for cobalt-60 tomotherapy. *Radiother. Oncol.*, **84,** p. S21.

20. Chng, N. (2005). An Investigation of inverse planning techniques for an experimental tomotherapy unit using a cobalt-60 source, MSc Thesis, Department of Physics, Queen's University, Kingston, Ontario, Canada.
21. Dhanesar, S., Darko, J., and Schreiner, L. J. (2012). Aperture superposition dose model versus pencil beam superposition dose model for a finite size cobalt-60 source for tomotherapy deliveries. *Med Phys.*, **39**, pp. 206–213.
22. Dhanesar, S., Darko, J., Joshi, C., Kerr, A., and Schreiner, L. J. (2013). Cobalt-60 tomotherapy: Clinical treatment planning and phantom dose delivery studies. *Med. Phys.*, **40**, p. 081710.
23. Dhanesar, S. (2013). The role of cobalt-60 source in intensity modulated radiation therapy: From modeling finite sources to treatment planning and conformal dose delivery. Thesis (Ph.D, Physics, Engineering Physics and Astronomy), Queen's University, Kingston, Ontario, Canada.
24. Cedric, X. Y., Shao, X., Zhang, J., Regine, W., Zheng, M., Ying, S. Y., Deng, J. and Duan, Z., (2013). GammaPod—A new device dedicated for stereotactic radiotherapy of breast cancer. *Med. Phy.*, **40**(5), p. 051703.
25. Rodin, D., Grover, S., Xu, M. J., Hanna, T. P., Olson, R., Schreiner, L. J., Munshi, A., Mornex, F., Palma, D. and Gaspar, L. E., (2016). Radiotherapeutic Management of Non-Small Cell Lung Cancer in the Minimal Resource Setting. *J. Thor. Oncol.*, **11**(1), pp. 21–29.

Chapter 3

Ferrous Sulfate–Benzoic Acid–Xylenol Orange Chemical Dosimetry System in Radiotherapy

Manoj K. Semwal[a] and Pandit B. Vidyasagar[b,c]

[a]*Department of Radiation Oncology, Army Hospital (R&R), Delhi Cantt, New Delhi, India*
[b]*SNT Marathwada University, Nanded, Maharashtra, India*
[c]*Department of Physics, University of Pune, Pune, Maharashtra, India*

manojsemwal@yahoo.co.in

The ferrous sulfate-benzoic acid–xylenol orange (FBX) dosimetry system in aqueous form has found utility in several types of applications with photon- and electron-based radiotherapy. The optical density (OD) measurements for dose estimation in the FBX were earlier performed with spectrophotometers. However, later it has been observed that simple filter-based colorimeters are equally effective for OD measurements for many applications. This reduces the cost of the FBX system significantly. FBX has also shown potential for applications in heavy-particle radiotherapy that is increasingly being used for cancer treatment. As a true 3 D dosimeter for modern radiotherapy delivery systems FBX in gel form is being extensively researched. FBX gel system could be very close to an ideal dosimetry system for 3D dose mapping in 3D-conformal radiotherapy, stereotactic radiosurgery/radiotherapy,

Radiation in Medicine and Biology
Edited by Pandit B. Vidyasagar, Sagar S. Jagtap, and Omprakash Yemul
Copyright © 2017 Pan Stanford Publishing Pte. Ltd.
ISBN 978-981-4745-92-5 (Hardcover), 978-1-315-20656-1 (eBook)
www.panstanford.com

intensity-modulated, and heavy-particle radiotherapy in terms of its performance and affordability.

3.1 Introduction

Radiation dosimetry has played an important role in standardizing and developing radiotherapy and other radiation applications. Accurate dosimetry for clinical and biological irradiators is critical for expecting the desired outcomes. From gas-filled ion-chambers to various solid-sate detectors such as thermoluminescence detectors (TLDs), diodes, metal oxide semiconductor field effect transistors (MOSFETS) and films have been used for dose measurement purposes. All these dosimeters have their own specific advantages for given situations. The technology for the delivery of radiation, especially in radiation oncology, has evolved over the century from the early uniform two-dimensional (2D) delivery of radiation to today's 3D conformal radiation delivery technology using modern linear accelerators with intensity-modulated radiotherapy (IMRT), image guided radiotherapy (IGRT), stereotactic radiosurgery (SRS), and stereotactic body radiotherapy (SBRT) facilities. In addition, charged particles especially proton facilities are steadily coming into clinics. Brachytherapy has also advanced through improvements in delivery systems and 3D image–based planning and source dwell position optimization. All these developments in dose delivery technology and 3D image–based planning have resulted in high dose conformity around the tumor volume and sharp dose fall-off beyond it. The current, mainly morphological imaging, for tumor delineation is steadily giving way to better functional and molecular imaging leading to biologically guided radiotherapy (BGRT). This has the potential of changing the current paradigm of uniform irradiation of the target volume to non-uniform irradiation (dose painting) depending on the biological property of the various regions of the target. Accurate dosimetry for medical and biological irradiators is, therefore, critical for expecting the desired outcomes.

The main challenge in day-to-day clinical implementation of these high delivery technologies is their extremely low tolerance for errors. The cost of mistreatment due to even small errors is substantially high in terms of treatment outcomes/toxicities with these technologies [3]. Therefore, verification before, during and after delivery of radiation becomes important to ensure accurate delivery of treatment. Following are some of the dosimeters available for dosimetry verification as part of the quality assurance program:

3.1.1 Chemical Dosimetry

Measurement of chemical changes brought about by ionizing radiation in a system for quantifying the radiation dose is considered as chemical dosimetry. Many materials have been experimented with for their suitability as chemical dosimeters.

3.1.2 Fricke System

The Fricke system is possibly the oldest and the most reliable chemical dosimeter. The oxidation of ferrous ions in acidic aerated solutions was proposed as a chemical dosimeter by Fricke and Morse in 1927 [7]. The dosimeter is capable of achieving an accuracy of better than 3% with a measuring range 40 to 400 Gy. On exposing the solution to radiation, the oxidation of ferrous ions to ferric ions is dependent on the absorbed dose. The ferric ion concentration may be measured in a variety of ways such as chemical titrations, the most reliable one employing a spectrophotometer. The measured difference in absorbance is related to the absorbed dose in the solution according to the following equation based on the Beer–Lambert law:

$$D = \frac{N_{\mathrm{A}} \times \Delta(\mathrm{OD})}{\rho G T \varepsilon_{\mathrm{m}}},$$

where D is the average absorbed dose in the solution, N_{A} the Avogadro constant, Δ (OD) the difference in absorbance before and after irradiation, G the ferric ion G-value, t the cell path length, and ε_{m} the molar absorption coefficient for ferric ions.

The ferric ions have absorption maxima at the UV wave lengths of 224 and 304 nm and the reported and recommended molar absorption coefficients values are ε_{224} (Fe^{3+}) = 4565 M^{-1} cm^{-1} and ε_{304} (Fe^{3+}) = 2205 M^{-1} cm^{-1}. Thus the dosimeter sensitivity is approximately doubled when measurements are carried out at 224 nm. However, impurities from plastic containers are more troublesome at 224 nm than at 304 nm [19]. The response of the dosimeter is linear up to 350 Gy. The Fricke system has been extensively studied over the years using photons and electrons. The ICRU in its report No. 14 [19] recommended it for absorbed dose standardization. At present it is one of the three methods, along with calorimetry and ionometry, that are sufficiently accurate to form the basis of primary standards for measurements of absorbed dose to water [18]. The Fricke primary standard for absorbed dose to water is based on the response of the system to an electron beam of known energy and fluence completely stopped in the Fricke solution [6, 37].

3.1.3 Ferrous Sulfate–Benzoic Acid–Xylenol Orange Dosimetry System

The ferrous sulfate–benzoic acid–xylenol orange dosimeter is a modified Fricke dosimeter developed by Gupta and co-workers [8–10, 12–14] to increase the sensitivity of the latter at low doses. The measurable range of the FBX dosimeter is from 0.1 to 3000 cGy. The aerated FBX system contains 0.20 mM (0.2 mol m^{-3}) ferrous ammonium sulfate, 5.0 mM (5.0 mol m^{-3}) benzoic acid and 0.20 mM (0.2 mol m^{-3}) xylenol orange in 0.08N (40 mol m^{-3}) sulfuric acid (FBX dosimeter). Originally the sulfuric acid concentration was 0.025 mol m^{-3} [10]. However, it was found out later that the acid concentration for optimal complex formation depended on the supply source of XO, for example for Loba Chemie XO it was 0.08N [11]. The optical density measurements are done at about 540 nm in the visible range unlike UV range of the Fricke system, thus simplifying the measurements as well. The low dose measuring capability of the FBX system opened new opportunities for its applications in radiotherapy.

3.1.3.1 Applications of FBX dosimetry in radiotherapy

The aforementioned properties of the FBX system show that it not only retains all the advantages of the Fricke system, it has the added advantages such as the low dose measuring capability, simplicity of measurements in that spectrophotometric measurements are performed in the visible range. Also the molar absorption coefficient of Ferric-XO complex is independent of temperature of the solution in the normal room temperature range unlike Fricke [10]. The FBX system, therefore, can be a very versatile and cost effective dosimetry system in radiotherapy. It has indeed shown utility in dose measurements in a water medium for photon and electron beams of various energies used in radiotherapy. It has been found useful for PDD measurements, output calibration [15, 16], postal dose intercomparison [12] in external beam therapy and for source calibration [25] and in vivo dose measurements [4] in brachytherapy. Pejuan and Kuhn [31] found its potential for neutron dosimetry in radiation protection. Benedetto and Harris [2] measured absorbed dose from radionuclide solutions by mixing FBX with them. Gupta and Nilekani [17] measured electron (beta) backscatter from a ^{90}Sr–^{90}Y source by using thin layer of the FBX solution. They found out that the backscatter was underestimated by the other dosimetry techniques such as ionometry and TLD method.

3.1.3.1.1 In vivo (entrance) dose measurements with the FBX system

In radiotherapy, the aim is to deliver a known radiation dose with accuracy to a target volume in a patient and spare the surrounding normal tissues, in order to eliminate the cancer cells with least side effects of radiation. However, in practice it is difficult to meet the goal perfectly. There is inaccuracy or uncertainty in the dose delivered to the target and at the same time surrounding normal tissues also receive part of the radiation dose aimed at the target. Among the many QA tests performed for ensuring accurate delivery of radiation dose, in vivo dosimetry provides an overall test of the whole system up to and including patient setup and treatment delivery. It can consist of entrance dose, exit dose, intracavitary dose measurements, and the determination of the dose delivered to critical organs such as the eyes or the gonads.

In vivo dosimetry has a long history. Sievert performed patient dose measurements with small ion chambers in 1932 [23]. Thermoluminiscent dosimeters (TLD) for in vivo dosimetry were introduced in routine therapy in the 1960s and since then have been used widely for the purpose [5, 21]. Among the many TL materials, LiF has been the most extensively used TL compound in radiotherapy dosimetry [26]. When considerable care is taken, precision of approximately 3% may be obtained with TLDs.

Following the work of Rikner [33], the semiconductor diodes have been increasingly used for in vivo dose measurements for the last two decades [1, 5, 22–24, 27]. The diodes have the advantages of instant read out which can help in correcting the source of error immediately and better spatial resolution because of small measuring volume. But diode response is more energy dependent than the TLDs. MOSFET dosimeters due to their very small size and ruggedness are also considered useful for in vivo dosimetry [32].

The FBX system could be a suitable dosimeter for in vivo measurements owing to its properties such as water equivalent composition, energy and temperature independent response. Semwal et al. [35] carried out in vivo dose measurements with FBX dosimeter for patients undergoing external beam radiotherapy with telecobalt machines. Dynamic dose profile measurements (virtual wedge) were also carried out with the FBX system [34]. Carbon beam dosimetry using FBX showed that the sensitivity of the dosimeter is about 25% that with gamma or x-rays [36].

3.1.3.1.2 Assessing potential of FBX dosimeter for dynamic (virtual) wedge profile determination

The dynamic wedge (DW) was originally proposed by Kijewski and Levene in 1978. The advents of digital linear accelerators have made it possible to use DW in clinics. The first implementation of dynamic wedge was accomplished by Leavitt in 1991. It has now become a common feature of the modern linear accelerators and is preferred over the physical wedges that have been in use for a long time. A wedge-shaped profile by dynamic wedge technique is generated with computer controlled dose delivery rate and jaw movement. The monitor unit output from the treatment machine as a function of moving jaw position is specified by a segmented treatment table stored in the accelerator computer. In effect, the dynamic wedge is created by delivering an open field radiation for most of the beam-on duration. For a small part, the radiation is delivered while one of the independent jaws of the collimator is continuously moving. The jaw may either be opening-out or closing-in depending upon the design chosen by a particular manufacturer of a linear accelerator.

The dynamic wedge has led to significant practical and dosimetric improvements over the physical wedge (PW) such as

(i) no beam hardening effect as there is no filtration of the beam; (ii) saving in beam-on and setup times; (iii) improvement in dose homogeneity within the target volume; this is possible because unlike physical wedges, the isodose profile angle is maintained at deep depths in the DW case, whereas in the PW case the profiles flatten in the "toe" region of the wedge at deep depths; and (iv) reduction in the surface and peripheral dose.

The percentage depth dose measurements for an open or for a DW field are not different and a single detector is enough for the purpose. However, to measure the dynamic wedge profiles in directions perpendicular to the beam axis, a dosimeter has to be static in order to integrate dose during the entire dose delivery process of one DW operation. Multiple diode detector/ ion chamber arrays, called as linear detector arrays (LDAs) and films are commonly used for such measurements, having their own advantages and disadvantages. One of the difficulties with the LDAs is that each of the two dozen odd detectors needs to be calibrated prior to such measurements. The films may be used more efficiently as it involves less measurement preparation and fewer times of DW operations as compared to the LDAs. However, many steps in the film dosimetry procedure such as film processing are very sensitive to subjective errors and hence require utmost care for reliable dosimetry measurements.

The FBX dosimeter has not been studied for DW profile measurements. The DW is a precursor to dynamic conformal radiotherapy and suitability of FBX for the dosimetry of DW would open up further opportunities for its applications in the latest field of intensity-modulated radiotherapy. In addition to the qualities of the FBX dosimeter mentioned in the preceding paragraphs, it also needs to be emphasized that there is insignificant additional cost in making an array of dosimeters required for dynamic wedge dosimetry. The low cost of the FBX dosimeter was also a factor in undertaking this study.

3.1.3.1.3 Response of the FBX dosimeter to a carbon beam

The linear energy transfer (LET) value of a radiation is an important consideration in chemical dosimetry as the response of the dosimeter depends on the LET. For low-LET radiation such as x-rays, gamma rays, and electrons, the free radical (H, OH) yields are high and the molecular yields (H_2, H_2O_2) are low since the concentration of the free radicals is low along the tracks of such radiation and hence relatively few radicals react with one another while the majority react with solutes. The water radiolysis yield (*G*-values) for cobalt-60 gamma rays of these primary products in 0.4 mol/L sulfuric acid is as follows: G_{H} (3.70), G_{OH} (2.92), G_{H2} (0.39), G_{H2O2} (0.78), where as the *G*-values for $\mathrm{B}^{10}(n, \alpha)\mathrm{Li}^7$ are G_{H} (0.23), G_{OH} (0.41), G_{H2} (1.66), G_{H2O2} (1.57) [38]. Many studies are available on the response of the Fricke system to various types of charged particle beams having a wide range of LET values. Chatterjee and Magee (1980) developed a successful model for heavy-particle tracks and calculated the response of the Fricke system for H, He, C, Ne, Ar, and Fm particles. Laverne and Schuler (1987) experimentally determined the response of Fricke system for He, Li, Be, B, and carbon beams with energies up to 35 MeV. The LET values are 20 KeV/μ for the helium and 100 KeV/μ for the carbon beam. These studies showed that the response of the Fricke system decreases with increasing LET. In a review article, Laverne (2000) has given an account as to how higher LET values result in lower *G*-value for a system like Fricke.

Protons and heavy particles such as helium ion, π-meson, carbon, neon, oxygen, etc. are considered to be useful in radiotherapy, some of them being in use also, for their superior physical and biological characteristics. Among these particles, the optimal mass suitable for radiotherapy is considered to be that of carbon or oxygen. However, work has not been reported on the response of the FBX system to high LET particles except for $^{10}\mathrm{B}(n, \alpha)^7\mathrm{Li}$ recoils where the sensitivity of the FBX has been found to be 22.1% as compared to that for gamma rays.

To assess the potential of the FBX system in charge particle radiotherapy, the logical first step would be to determine the response of the system to a charge particle beam. Carbon beam has been chosen in the present work for its applicability in radiotherapy.

3.2 Experiments

3.2.1 Preparation of FBX Dosimetry Solution

The first-time cleaning of the glassware (5 mL beakers, 500 mL volumetric flask, 0.2 mL pipette), plasticware (wash bottles, irradiation tubes, etc.), and stainless steel spatula was done as per the following standard procedure: Glassware was kept filled with 1:1 HNO_3 and H_2SO_4 for 24 h and then washed with running tap water and finally rinsed with distilled water. Plasticware were kept filled or immersed in 10% HNO_3 for 24 h and then washed with running tap water and finally with distilled water. The irradiation tubes were filled with distilled water or dosimetric solution and irradiated with a dose of 50 Gy when used for the first time. The stainless steel spatulas were cleaned in running tap water and then distilled water. When not in use, the irradiation tubes were kept filled with dosimetric solution for longer duration and with distilled water for shorter duration.

First, distilled de-ionized water (DW) easily available in a hospital setup, instead of double distilled water, was used for solution preparation. To prepare 500 mL of FBX dosimetric solution, standard method mentioned in literature was followed [15]: To get the composition of 2×10^{-4} mol dm^{-3} (0.20 mM) ferrous ammonium sulfate (FAS), 2×10^{-4} mol dm^{-3} (0.20 mM) xylenol orange (XO) and 5×10^{-3} mol dm^{-3} (5.0 mM) benzoic acid (BA) in 0.04 mol dm^{-3} (0.08N) sulfuric acid (SA), 39.2 mg FAS, 76.1 mg XO and 305.3 mg BA were required. Weighing of these compounds in a microbalance was done in 5 mL glass beakers instead of using foils/ paper slips to avoid contamination.

The BA was dissolved in DW in a 500 mL volumetric flask by warming it in a water bath. The solution in the flask was now allowed to cool or was force cooled under running tap water to bring it to room temperature. Using a 0.2 mL pipette, 1.066 mL of concentrated SA was added to the solution. The FAS and XO were added to it and the volume was made up to 500 mL by adding DW. To aerate the solution sufficiently, the stopper cap of the volumetric flask was put on tightly and the solution shaken by turning the flask upside-down a number of times. The stopper was removed to let some air into the flask and placed back onto the flask. The process was repeated five times to aerate the solution. The solution thus prepared was stored in a refrigerator.

From the initial observations and results, it was realized that the method of solution preparation starting from weighing the chemicals to dissolution of the constituents needed streamlining to get reproducible results. Some of the precautions taken were (i) addition of FAS in the solution only after the addition of sulfuric acid. This was done to avoid formation of precipitate during solution preparation the cause for which was the formation of hydroxides by ferrous ions in water in a non-acidic medium. (ii) in no case was FAS added to heated solution to avoid thermal oxidation of ferrous into ferric (iii) excess chemicals once taken out into measuring beakers during weighing, were thrown away and not returned to the respective containers, to avoid contamination.

3.2.2 Spectrophotometer, Colorimeter and Optical Density Measurements

For most of the work reported in this chapter, spectrophotometer used was Spectronic 20D from Milton Roy and Co. and colorimeter was Photochem Micro from Aimil Ltd., unless specified otherwise. The spectrophotometer was used with the aim to assess the performance of the colorimeter-based FBX dosimeter. The absorbance path length in the spectrophotometer was 1.2 cm and that in the colorimeter 1.0 cm. The Spectronic 20D had the spectral slit width 20 nm and wavelength readability 1 nm. The colorimeter had a set of five monochromatic filters having peak transmissions at 420, 490, 540, 590 and 650 nm, respectively. Both the absorption measuring devices were single beam type and had circular aperture cuvette holders and test-tube like cuvettes. It was observed that changing the orientation of the cuvettes in the holders by rotating the cuvettes affected the absorbance value displayed. Therefore, orientation marks with indelible ink were made on the cuvettes and the holders for reproducibility of the positioning. Cuvette to cuvette variation was also noticed in both the cases. To avoid its influence on the experimental reading, cuvettes were carefully selected by matching absorbance for distilled water and also for FBX solution.

The wavelength calibration and linearity check of the spectrophotometer were done as follows:

A solution of cobalt chloride using 2.25 gm of the compound to make 100 mL of solution was prepared in 1% HCl. For wavelength calibration, a graph between wavelength and OD was plotted to find out the wavelength of maximum OD (absorbance). As shown in Fig. 3.1, the maximum OD value occurs at about 510 nm, which agrees with the expected value for cobalt chloride.

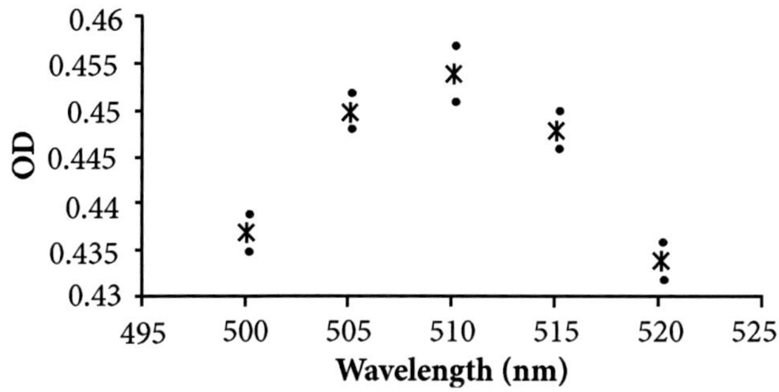

Figure 3.1 Wavelength (nm) vs. average OD. Measurements on spectronic 20D for a cobalt chloride solution. (:) shows the standard deviation (1SD).

To check the linearity, the concentration of the solution was varied in steps by making it half, one fourth and so on and the corresponding absorbance value at 510 nm was noted at each concentration. The straight line graph in Fig. 3.2 shows the linear response of the Spectronic 20D. While measuring the samples that were stored in refrigerator after irradiation, we noticed that optical density values shown were not stable.

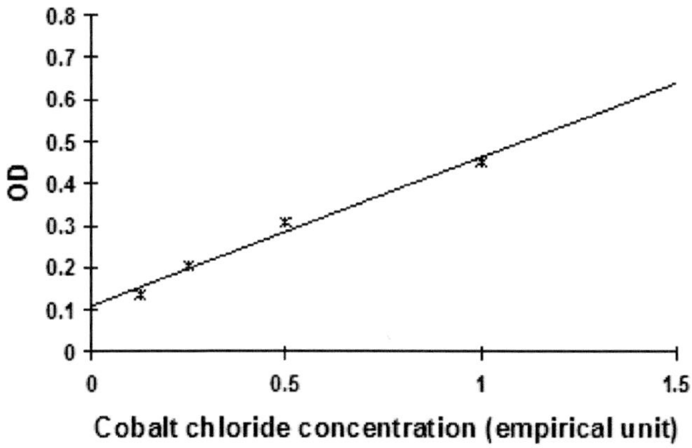

Figure 3.2 Linearity test of spectronic 20D. Cobalt chloride solution concentration vs. average OD values. Measured on spectronic 20D at 510 nm.

It was found out that for the samples taken from the refrigerator, condensation on the outer surface of the cuvettes was responsible for the changing OD values. To avoid this, it was decided to allow the samples to attain room temperature before proceeding with the OD measurements. It was also a standard practice to rinse the cuvette twice before every sample was read-out, with the blank (stock) solution. The colorimeter or the spectrophotometer was zeroed with blank solution after reading every two samples to correct for any drift.

3.2.3 Determination of Optimum acid Concentration and Maximum Absorption Wavelength

The ferric-xylenol complex formation is optimum at a particular sulfuric acid (SA) concentration in the solution. The optimum acid concentration and the wavelength of maximum absorption values depend, to some extent, on the maker of chemicals particularly that of XO, and the absorbance measurement setup. To determine the optimum SA concentration (in terms of normality) for the chemicals used, FBX solutions of various normalities in respect of SA were prepared and their response measured by irradiating them to a fixed dose value of 200 cGy on a cobalt teletherapy machine. The graph in Fig. 3.3 shows the response of FBX at various normalities.

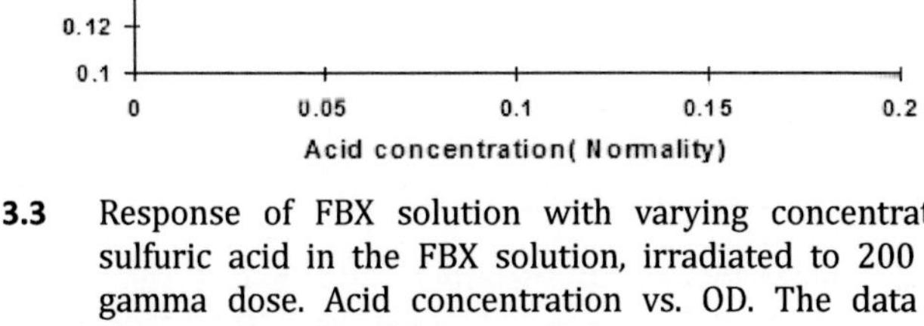

Figure 3.3 Response of FBX solution with varying concentration of sulfuric acid in the FBX solution, irradiated to 200 cGy of gamma dose. Acid concentration vs. OD. The data points show average OD of three readings.

The OD value is maximum for 0.08N solution, which means that complex formation is optimum at this normality. The value obtained agrees with that reported by Gupta et al. (1994) [11].

The wavelength of maximum absorption (λ_{max}) as measured on three makes of spectrophotometers (Spectronic 20D, Hitachi 3300, Hitachi U2000) for the FBX varied between 544 to 555 nm for chemicals of different brands. For the chemicals that we finally decided to use namely, FAS, BA, and SA from E-Merck, Germany and XO from Loba Chemie, Austria, the λ_{max} value was 548 nm for the Spectronic 20D (Fig. 3.4).

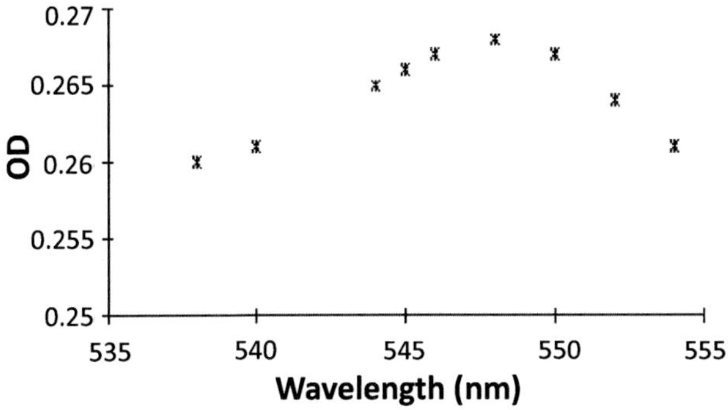

Figure 3.4 Determining the wavelength of maximum absorption for the FBX system. Wavelength vs. OD for FBX solution irradiated to 200 cGy of gamma dose, measured on Spectronic 20D.

In the case of the colorimeter, the maximum absorbance was about the 540 nm band. Based on the above observations, it was decided to use λ_{max} = 548 nm for the work reported in this thesis in the case of a spectrophotometer and 540 nm in the case of a colorimeter.

3.2.4 Suitability of Polypropylene Tubes for Irradiation

The polypropylene tubes (PTs) procured from the market of 13 mm outer diameter with approximately 1 mm wall thickness and 55 mm height were used for in vivo dose measurements. The tubes were given prior treatment as mentioned earlier before being used for experiments. The tubes were irradiated with specially made Perspex buildup caps and without the caps and the response of the FBX system was measured at various doses. The buildup cap wall thickness was approximately 3 mm. The data presented in Fig. 3.5 show that the response of the FBX system is independent of the buildup cap.

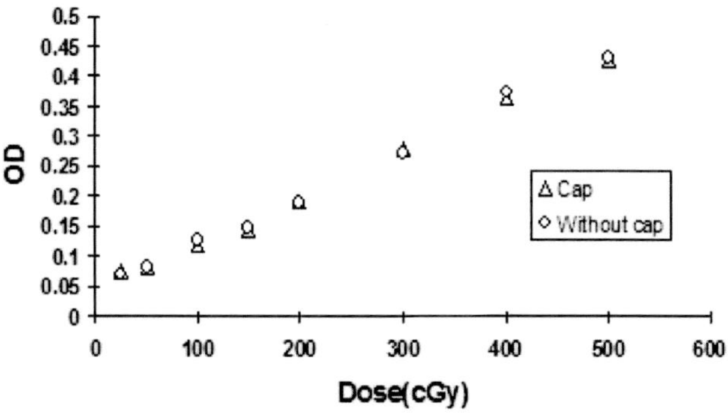

Figure 3.5 Dose vs. OD for FBX dosimeter irradiated with cobalt-60 gamma rays. The FBX solution was filled in polypropylene tubes used with and without Perspex buildup cap. The OD values shown are average for five tubes at each dose level.

The reason could be that the solution thickness in the tubes along with the 1 mm wall thickness of the tubes was sufficient to provide the necessary buildup at cobalt-60 gamma energy (1.25 MeV). Therefore the tubes were always used without buildup cap for better spatial resolution and convenience.

3.2.5 Minimum Measurable Dose with Colorimeter and Spectrophotometer

To assess the suitability of a colorimeter-based FBX dosimeter, one of the preliminary requirements felt was to determine the minimum measurable dose with reasonable accuracy (±3% to 4%) in the routine setup. The lower the measurable dose, the better is the dosimeter. Reproducibility of the FBX dosimeter in a hospital environment was another important factor that needed to be determined before undertaking the main dosimetry work.

Following the aforementioned procedures, it was decided to find out the reproducibility of the dosimeter response for the existing setup. Five fresh batches of solution prepared over a period of three months were tested for reproducibility. Equal dose (100 cGy) under identical geometric setup was delivered to five samples from each batch. The filling of solution into PTs, the irradiation of the tubes and the OD measurements were carried out on the day the solution was prepared for each batch. The effect of storage of the FBX solution in the glass container, usually the measuring flask in which the solution was prepared, was also carried out. The dosimeters (filled PTs) were prepared from the FBX solution on the day of solution preparation and then on 7th, 15th, and 21st day. The reproducibility of the dosimeters was within ±3%, which was in close agreement with the published value [15]. However, it was observed that storage of dosimetric solution in the irradiation tubes either before irradiation or after irradiation resulted in larger deviations. Deviation as high as 6% was noticed for dosimeters read after 2 days. This could be due to reaction of the dosimetric solution with the tube material or the evaporation of the dosimetric solution from the tubes. Therefore, it was decided to fill, irradiate, and read the dosimeters on the same day.

The FBX dosimeter was then calibrated for the dose range of interest in the prevailing environmental condition. For calibration, FBX solution was filled in the PTs and the irradiation was carried out on the TH780E cobalt teletherapy machine for doses varying from 5 to 200 cGy. The teletherapy machine output was measured with an ion-chamber–based secondary standard dosimeter whose calibration was traceable to the Secondary Standards Laboratory, BARC, Mumbai. Three tube samples were irradiated at each dose level. To minimize timer error at the lower doses, the exposure time was increased by irradiating the tubes at longer distances (up to 170 cm SSD). A water phantom was used for irradiation purposes. Absorbance measurements were then carried out on the Spectronic 20D spectrophotometer at 548 nm and on the colorimeter at 540 nm. To minimize contamination, the tubes were always rinsed with distilled water and then three times with FBX solution before being filled for irradiation and measurements. It was ensured that the OD was measured at least 30 min after irradiation to allow for the reactions leading to ferric-xylenol complex formation to get completed. The dose Vs OD graphs (Fig. 3.6) show that the slope of the graph is more for the spectrophotometer as compared to that of colorimeter implying the higher sensitivity of the spectrophotometer.

It could be inferred from the data that the minimum measurable doses with reasonable accuracy with the colorimeter and the spectrophotometer are approximately 5–10 and 5–7 cGy, respectively. This performance was considered adequate to use the dosimeter for the proposed work. However, apart from the sensitivity factor, a limitation with the colorimeter was that it required a minimum of about 1 mL solution for OD measurements and hence, as it would be observed later, the colorimeter could not be used in situations where the solution quantity available was less than 1 mL.

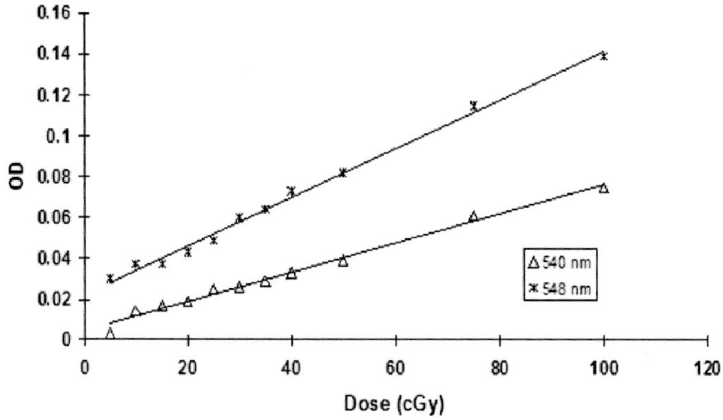

Figure 3.6 Dose vs. average OD values for FBX dosimeter irradiated on a telecobalt machine in polypropylene tubes and OD measured with spectrophotometer (548 nm) and colorimeter (540 nm).

3.2.6 In vivo (Entrance) Dose Measurements with the FBX System

The FBX solution was prepared as per the method described in Section 3.2.1. Stoppered cylindrical polypropylene tubes, details mentioned in Section 3.2.4, were used as dosimeters (irradiation vials). It was planned to measure entrance doses on patients treated on a TH-780E cobalt teletherapy machine. A total of 220 setups (143 fields, 100 patients) were considered in this study. Usually one but occasionally two measurements (setups) were done per field. We selected only two simple categories of patients for our work namely patients getting radiation in the pelvic and in the cranial regions, with parallel opposed fields thus avoiding uneven surface for placement of the irradiation tubes. No wedged or oblique beams were chosen for this work as the primary aim had been to assess the suitability of the FBX system for dose measurements on a routine basis in a hospital environment. Of the total 220 setups measured, 100 were cranial irradiations and 120 pelvic irradiations, both categories having 50 patients each. All except 6 out of the 100 patients were treated isocentrically by parallel opposed beam portals. The remaining 6 were treated with constant SSD technique. For each patient, entrance dose (dose at d_{max}) was measured at the central axis of the beam.

The calibration of the FBX system (dose versus OD) for entrance dose measurement was done on the day a new solution was prepared. For calibration, the FBX filled tubes were placed on the side surface of an open top water phantom. The wall thickness was particularly thin (about 2 mm) over an area of about 15 cm × 15 cm from where the beam entered the surface simulating patient geometry (Fig. 3.7).

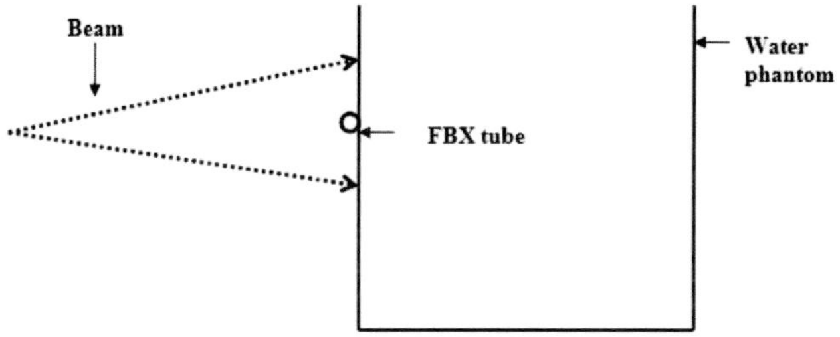

Figure 3.7 Schematic diagram of FBX dosimeter entrance dose calibration setup. The radiation beam from the telecobalt machine fell on the one of the walls of the open top water phantom.

The machine output was measured in the same geometry at 5 cm depth with a 0.6 cc farmer-type ion chamber along with Unidose, secondary standard dosimeter (both from PTW, Germany) for a 10 cm × 10 cm field size at standard SSD (80 cm) for the cobalt machine. The output at 0.5 cm depth (d_{max}) for cobalt-60 gamma rays was calculated from this using percentage depth dose value. As has been described in Section 3.5, the irradiation tubes' dimension was sufficient to provide necessary buildup for cobalt-60 gamma rays and hence no additional buildup was used with the tubes. The known radiation doses were delivered at $d_{max\ position}$ and the response of the FBX dosimeters noted in terms of dose versus OD. The calibration graph of the FBX dosimeters in this geometry was used to determine the entrance dose (dose at d_{max}) in the patients from the FBX response on the patients. A typical calibration graph is shown in Fig. 3.2. The calibration procedure was repeated every time a new solution was prepared. The solution prepared was stored in the measuring flask itself, to avoid contamination by using a different storage container, in a refrigerator at about 5°C. A solution once prepared was not used for more than one week in order to avoid the errors due to thermal oxidation, however small they might be. Filling of the tubes, their irradiation, and the OD measurements on the colorimeter (Photochem micro 5 filters) at 540 nm were always completed on the same day. The patient setups were performed as usual by the technicians and after completion, the FBX tubes were fixed on the skin of the patients in the centre of the irradiation field. Expected entrance doses were calculated manually from Tissue air ratio (TAR) or percentage depth dose (PDD) data.

To assess the accuracy with which the FBX dosimeter can be used for entrance dose measurements under clinical conditions, two patients were monitored over 8 treatment fractions, one for cranial irradiation and the other for pelvic irradiation.

3.2.7 Assessing Potential of FBX Dosimeter for Dynamic (Virtual) Wedge Profile Determination

The FBX solution was prepared as per the method described in Section 3.3.1. To measure the dynamic wedge profiles, an array of dosimeters was needed which remained static during the entire wedge operation.

The conventional LDA setup was imitated for measurement. Several FBX dosimeters (FBX solution filled tubes) were used to form an array, the number of tubes required depending of the tube size, spacing arrangement and the irradiation field size. Three sizes of dosimeters were used in this study namely, (i) stoppered polypropylene tubes (PTs) of about 1.4 cm outer diameter, 5.5 cm length and 1 mm wall thickness available in the market; (ii) tubes made "in-house" from readily available needle covers of 5 mL disposable syringes; with some modifications, the needle bases were used as stoppers for the tubes; the outer diameter and length of these cylindrical needle cover tubes

(NCTs) are about 0.6 cm and 5 cm, respectively; and (iii) tubes made from IV drip catheter (IVTs), by heat sealing at both ends, with outer diameter and length of about 0.4 cm and 5.0 cm, respectively. Before using these irradiation tubes for experimental purpose, they were treated as follows: first washed under running tap water and then kept filled with 10% nitric acid for 24 h. After this, they were again rinsed with tap water followed by rinsing with distilled water and then filled with FBX solution and irradiated with a dose of approximately 5000 cGy. In the case of IVTs, the whole lengths of the IV catheters, before making individual tubes by heat sealing at both ends, were treated in this way. The filling of NCTs and IVTs with FBX solution was accomplished with a syringe to avoid formation of air spaces in them.

In this study, a Siemens Primus accelerator was used to deliver 6 MV photon beam with 45° dynamic wedge (virtual wedge in Siemens machines) for a field size of 10 cm × 10 cm at depths of 1.5 cm (d_{max} for 6 MV photons), 10 cm and 20 cm in a Wellhoffer Blue Phantom at SSD 100 cm. For comparison of the FBX, measured profiles films were used. The films (Kodak X-Omat V verification films) were irradiated in a solid water phantom (virtual water from Med Tech, USA) and read on a Wellhoffer WP102 film densitometer with a spatial resolution of 2 mm.

The tubes were placed horizontally, in a set of 13 or 15 tubes in the case of PTs, and 25 tubes in the case of NCTs on a Perspex rectangular slab, which in turn was mounted on the regular guide rail of the blue phantom. Double-sided sticking tape was used for placing the tubes on the slab properly. The tubes had their long axis perpendicular to the long axis of the Perspex slab (Figs. 3.8 and 3.9).

The collimator rotation was such that the moving jaws were aligned along the length of the Perspex slab. The IVTs (25 in number) were irradiated in a "virtual water" solid phantom keeping 0.5 cm jelly bolus (Med Tech) immediately above and below them to reduce the formation of air spaces between the IVTs (Fig. 3.10). Wet cotton was used to fill the air-spaces still remaining.

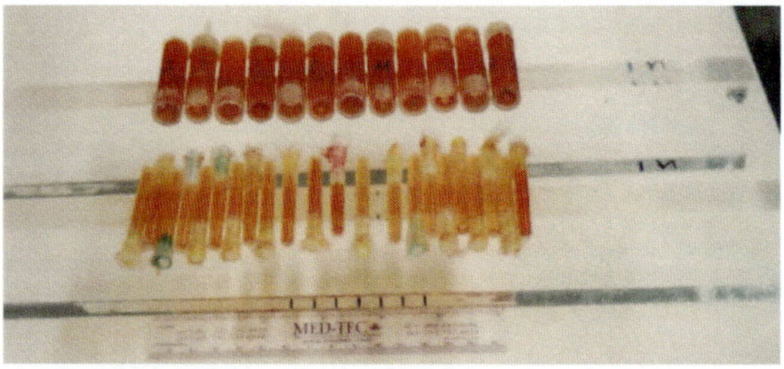

Figure 3.8 Polypropylene tubes (bigger ones) and needle cover tubes placed on Perspex slabs to form "arrays" ready for mounting in a radiation field analyzer (RFA).

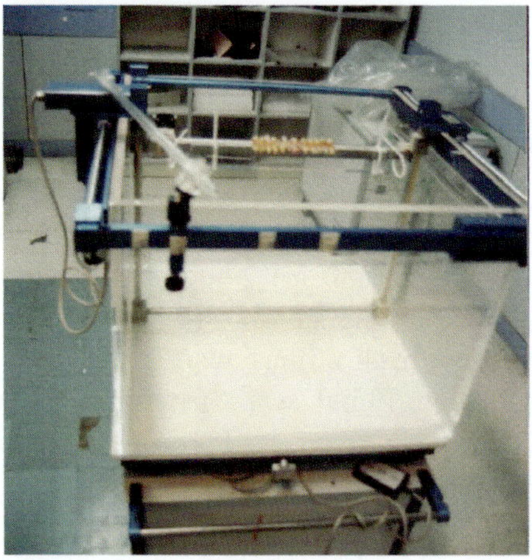

Figure 3.9 A needle cover tube (NCT) array mounted in an RFA guide rail for irradiation.

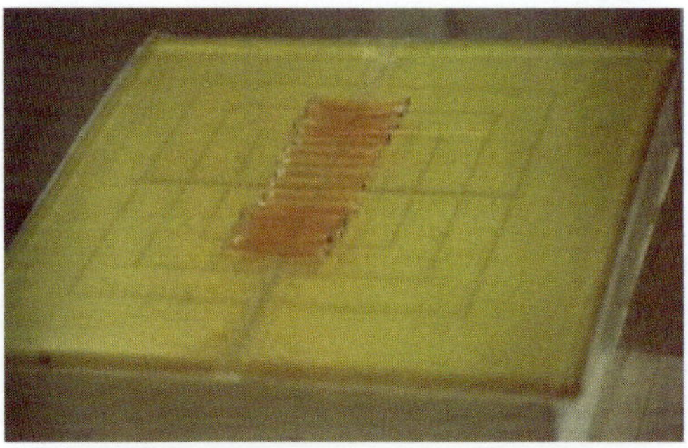

Figure 3.10 IV chamber tubes (IVTs) placed on jelly bolus on a solid water phantom for irradiation.

The PTs being larger in size, were placed without spacing between them, whereas in case of NCTs and IVTs, they were kept 1 cm apart up to 3 cm on both sides of the central axis, in the main part of the beam, and without spacing beyond this region to cover the steep dose gradient region. The final position of each tube with respect to the central tube was noted after the placement. The individual tube positioning error was 0.5 mm in the three sets of data taken with each type of "tube array." The dose delivered at the central axis at all the depths was 200 cGy. Each type of the irradiation tubes were filled irradiated and read on the same day. The optical density measurements were carried out on a colorimeter at 540 nm (Photochem micro 5 filters) for the PTs and the NCTs. For the IVTs, Hitachi U 3300 spectrophotometer having a 50 μL cuvette was used at 548 nm, the solution quantity available (about 200 μL) being insufficient for colorimetric measurements.

The films were irradiated in the solid water phantom at 100 cm SSD at 1.5 cm, 10 cm, and 20 cm depths. Films from the same batch were irradiated with different doses ranging from 5 cGy to 150 cGy at the depths of profile measurement to get the dose response curves of the film. Processing of the dynamic wedge films and the dose-response films was done at one time in an automatic film processor (Optimax from Protec, Germany) in order to avoid density changes due to altered processing conditions. Care was taken to use films from same batch for one set of experiment that is for dose–density calibration and for DW profile measurements. Ion-chamber (0.13 cc IC15) from Wellhoffer was also used to measure static open field profiles at 1.5 cm, 10 and cm depths and these profiles were compared with film profiles to validate the performance of the film. The off axis ratio values (OARs) were normalized to 100% at d_{max}. The percentage depth dose measured with the IC-15 chamber was used to re-normalize the relative dose profiles at various depths measured with the film. The optical density film profiles read on WP 102 film densitometer were converted into dose profiles with the help of dose vs. OD calibration curves at each depth. The OAR values corresponding to the tube positions were determined from the film profiles using WP700 software for comparison.

3.2.8 Response of the FBX Dosimeter to a Carbon Beam

The irradiation experiment was carried out at 15 UD Pelletron (16 MV Van de Graff machine) facility of the Nuclear Science Centre, New Delhi with 1 cm diameter carbon-12 beam. One of the improvisations required to carry out the experiment was to devise a technique for irradiation as the radiation beam from the Pelletron emerges horizontally. Further, the 55 MeV carbon beams that were available from the machine had a range of approximately 115 μm in water and hence the wall of container used for irradiation was required to be very thin to avoid attenuation to the beam in the wall itself. The LET for this beam in water as calculated by TRIM 98.0 Monte Carlo code varies from about 300 KeV/μm at the entrance to reaching a maximum value of about 800 KeV/ μm at 100 μm depth before the beam is completely stopped. Any irradiation vial to hold the FBX solution should have a wall thickness considerably less than 115 μm. A pilot experiment with 20 mL glass bottles was done to test the feasibility of the procedure. The sidewalls of the bottles had plane surfaces. For beam entry, a hole of approximately 2 cm diameter was cut on one of the sidewalls of the bottles. The hole was then covered with a 6 μm-thick polypropylene sheet with the help of glue. During irradiation, the solution was continuously stirred with a magnetic stirrer to homogenize the dose in the solution. The results obtained with this technique were not reproducible and also the dose and corresponding optical density values showed no relationship. Clearly the uncertainties associated with the experiment overshadowed the dose response pattern. We believe that among the many reasons for such an outcome, contamination of the solution with the glue used to paste the polypropylene sheet and non-homogeneous dose distribution in the solution owing to a large volume of the solution were the most important ones. Moreover, time taken to deliver doses relevant for our experiment (few cGy to few tens of cGy) to the solution was very long due to large volume of the solution and low particle flux (10^3–10^4 particles cm^{-2} s^{-1}) achievable from the Pelletron.

A new technique for irradiation in which very small quantities of solution could be homogeneously irradiated with minimum possibility of contamination of the solution was devised.

In this technique, a 6 micron thick polypropylene sheet was used to prepare 1.5 cm × 1.5 cm size sachets; heat sealed from all sides. Micropipette was used to inject known quantity of FBX solution into the sachets from a corner and the corner was once again heat sealed. Due care was taken during the filling to avoid air inside the sachets. For irradiation, the FBX filled sachet was mounted on a Perspex backing. It was kept stretched and compressed with a polypropylene sheet in order to make the solution thickness as uniform as possible. The sample thus mounted was kept at the exit window of the accelerator tube.

The energy of the carbon-12 beam falling on the sachet surface was 55 MeV, which was calculated to be 53.2 MeV when the beam entered the solution after traversing through the polypropylene sachet wall. The range of this beam in the FBX solution, which is practically water equivalent, was calculated to be approximately 115 μm. The calculations were based on Monte Carlo code TRIM 98.01 and were done at the Nuclear Science Centre, New Delhi. The total number of particles incident on a sample was found out from the monitor counts. The monitor system for counting was a surface barrier detector system calibrated prior to the sample irradiation and had an accuracy of better than ±2%.

The sachets were filled with 200 μL solution making the solution thickness to be approximately 900 μm. The samples were then irradiated with the carbon beam; total number of incident particles on the samples ranging from 2×10^6 to 4×10^7. The Irradiation time was approximately between 3 to 30 min. In all, six different particle exposures were used with three samples for each exposure. Since the carbon beam completely stopped in a sample, the value of beam energy and the total number of incident particles on a sample were used for conversion of particle exposure into absorbed dose in Gray, given the mass of the sample.

Gamma irradiation was done in stoppered polypropylene tubes of 3.5 cm length, 1 cm inner diameter and 1 mm wall thickness with the necessary Perspex buildup cap on a cobalt teletherapy machine. The cobalt machine output was measured with a secondary standard dosimeter whose calibration was traceable to the Secondary Standards Laboratory at Bhabha Atomic Research Centre, Mumbai. The uncertainty in the measurement is ±2%. The absorbance (OD) of irradiated samples could not be measured on the colorimeter since the available sample volume of about 200 μL was far less than the minimum required volume of over 1 mL for the colorimeter. Therefore the measurements were carried out on a Hitachi U3300 spectrophotometer at 548 nm against the unirradiated sample as blank. Each sample was measured at least 30 min after irradiation to allow the reactions to be completed.

3.3 Results and Discussions

Table 3.1 and Fig. 3.11 show the measured data for the cranial and the pelvic irradiations, respectively, in terms of expected and measured dose ratio (MD/ED). The results for all the 220 setups comprising both the categories of sites are shown in Fig. 3.11 in the form of a histogram. Similarly Figs. 3.11 and 3.12 show the results for individual categories, i.e., cranial and pelvic irradiation, respectively.

Table 3.1 Cranial irradiation data in terms of measured dose/expected dose (MD/ED %) for 100 setups ($N = 100$)

Sr. no.	MD/ED (%)	Sr. no.	MD/ED (%)	Sr. no.	MD/ED (%)
1.	95.3	35	99.1	68	101.8
2.	95.3	36	99.1	69	102.6
3.	96.2	37	99.5	70	102.6
4.	96.2	38	99.5	71	102.6
5.	96.8	39	99.1	72	102.6
6.	96.8	40	99.1	73	102.5
7.	97.0	41	100.9	74	102.5
8.	97.0	42	100.9	75	102.8
9.	97.1	43	100.8	76	102.8
10.	97.1	44	100.8	77	102.8
11.	97.3	45	100.8	78	102.8
12.	97.3	46	100.8	79	103.7
13.	97.3	47	100.8	80	103.7
14.	97.3	48	100.8	81	110.3

Sr. no.	MD/ED (%)	Sr. no.	MD/ED (%)	Sr. no.	MD/ED (%)
15.	97.8	49	100.8	82	110.3
16.	97.8	50	100.8	83	103.9
17.	98.1	51	100.4	84	103.9
18.	98.1	52	100.4	85	103.9
19.	98.3	53	100.5	86	103.9
20.	98.3	54	100.5	87	103.7
21.	98.2	55	100.7	88	103.7
22.	98.2	56	100.7	89	104.8
23.	98.9	57	101.9	90	104.8
24.	98.9	58	101.9	91	104.9
25.	98.1	59	101.7	92	104.9
26.	98.1	60	101.7	93	104.9
27.	99.0	61	101.7	94	104.9
28.	99.0	62	101.7	95	105.8
29.	99.2	63	101.9	96	105.8
30.	99.2	64	101.9	97	109.2
31.	99.2	65	101.9	98	109.2
32.	99.2	66	101.9	99	109.5
33.	99.3	67	101.8	100	109.5
34.	99.3	—	—	—	—

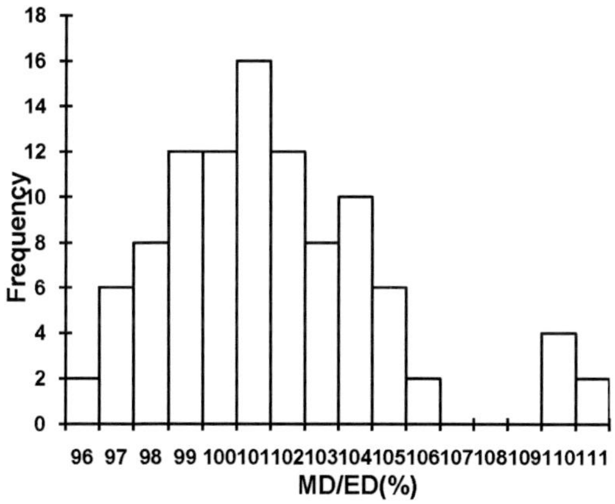

Figure 3.11 Frequency distribution of MD/ED (%) for cranial irradiation (N = 100), mean = 101.5%, SD = 3.3%.

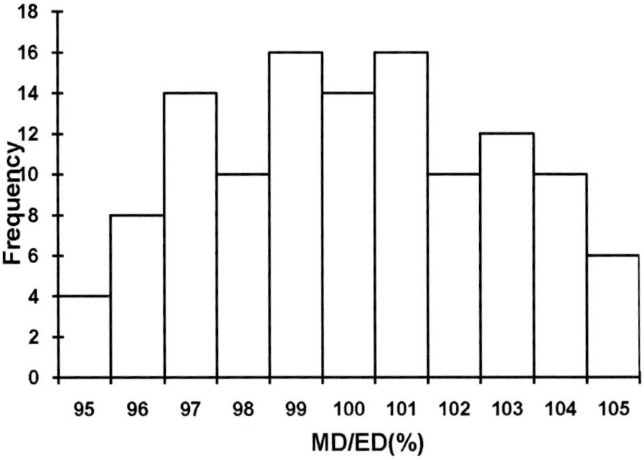

Figure 3.12 Frequency distribution of MD/ED (%) for pelvic irradiation (N = 120, mean 99.5%, SD 2.9%).

The histograms show the data as frequency distribution where (N) is the number of treatment setups measured, (X) is the mean value, and (SD) is the standard deviation.

The results for the total 220 setups of the entrance dose measurement showed a mean value of 100.2% and standard deviation of 3.2%. That means the average discrepancy or systematic error between the measured and the expected values was 0.2%.

In the case of whole-brain (cranial) irradiation, the mean value was 101.1% (N = 100, SD = 3.3%). Thus the average discrepancy was 1.1% on the positive side. That means the measured values were higher than the expected value. The maximum discrepancy was +10.3% and there were 8% (8/100) setups showing significant discrepancy (more than 5%). In the pelvic irradiation, the mean value was 99.5% (N = 120, SD = 2.9%). Therefore the average discrepancy or systematic error in pelvic irradiation was 0.5%. The maximum discrepancy in this case was −5.8% and the number of cases having significant discrepancy was 3.3% (4/120). The entrance dose reproducibility shows the daily variation of measured dose from calculated values in percentage for the two patients over eight days. It can be seen that the variation is within ±5% with a mean variation of 0.16 ± 2.8% for pelvis and 0.75 ± 3.1% for cranial irradiation, respectively. These values give an indication of the accuracy with which the

FBX dosimeter can be used to measure entrance doses on patients. For oblique beam incidences, wedge fields and other geometries the figures values may worsen.

As stated earlier, the overall discrepancy value of 0.2% was not significant. The reason could be the fact that all the procedure from measuring patient dimensions to calculating treatment times, etc. were done manually and hence chances of any systematic error creeping in the procedure were very rare. Leunens et al. [23] in a somewhat similar study but with semiconductor diodes have reported a systematic error of 2.75% and attributed it to inaccuracies in algorithm used for dose calculations with the treatment planning system. Figures 3.3–3.5 show that the overall results (N = 220) and the results for the two categories separately were near-Gaussian distributions. The pattern was similar to the published studies carried out by Leunens et al. [23] with semiconductor diodes.

The smaller discrepancy and SD in pelvic irradiation as compared to cranial irradiation was most probably due to reduced random errors in the former process. This could be attributed to the cross check and precision provided by the couch height scale of the cobalt machine in measuring the patient thickness in the AP and PA pelvic fields, which was not possible in the lateral fields used for brain irradiation. Hence, for brain irradiation one depended solely on the calipers measurements, which needed carefulness for precise measurements and thereby increased the chances of errors. Also the surface on which the irradiation tubes were placed was much more plane in the pelvic region as compared to skull region. Further, the chances of errors in positioning of the FBX dosimeters were higher in the cranial irradiation due to unevenness of the surface and also because the positioning was to be done in a vertical plane causing instability in the positioning. In cases where the head was not shaved, the problem was especially acute. The causes for the significant discrepancies could be ascertained in all the incidents. In eight of the total 12 cases, error in patient dimension measurement by more than 2 cm was the cause of the discrepancy. In two cases the discrepancy was noticed where treatment was planned for unequal beam weightage requiring different beam on times for the parallel opposed beams. The treatment was, however, erroneously delivered with equal beam on time. In two cases

(2/12), the constant source to surface distance (SSD) technique field was treated isocentrically as a constant source to axis distance (SAD) technique setup resulting in reduced SSD value for both the beams. This ultimately resulted in higher measured dose value as compared to expected dose.

It has been reported in literature that even with all precautions taken the film profiles still have errors up to 2%. Therefore before using the films for comparison with the FBX measurements, it was decided to validate the film measurements with ion-chamber measurements for at least two depths for open fields. The depths chosen were d_{max} (1.5 cm) and 10 cm. The d_{max} was included to observe whether the somewhat different beam spectrum at this depth will make difference in the film measurement accuracy. The ion-chamber measurements were regarded as definitive for testing the film accuracy. Table 3.2 shows the OAR values for the film and the ion chamber at 1.5 and 10 cm depths.

Table 3.2 Open field off axis ratio (OAR) values at two depths normal-ized at central axis (CAX) and measured with film and IC15

Off axis distance (cm)	OAR % (*d* = 1.5 cm) Film	IC15	OAR % (*d* = 10 cm) Film	IC15
7.0	—	—	8.5	9.3
6.8	4.9	5.3	8.8	10.1
6.6	5.1	5.6	9.5	11.1
6.4	5.5	6.0	10.6	12.5
6.2	6.0	6.7	12.5	14.7
6.0	6.7	7.5	15.8	18.4
5.8	7.8	9.1	21.7	24.7
5.6	10.0	12.5	32.1	34.7
5.4	14.5	18.6	43.2	48.3
5.2	27.6	30.2	58.2	63.2
5.0	45.8	47.7	79.5	76.0
4.8	61	64.7	85.9	84.8
4.6	81.8	80.9	92.6	91.0
4.4	92.1	90.5	—	—
4.3	—	—	96.8	94.8

Off axis distance (cm)	OAR % (d = 1.5 cm) Film IC15		OAR % (d = 10 cm) Film IC15	
4.2	97.4	95.7	—	—
4.0	99.9	98.1	98.5	96.4
3.8	101.1	99.3	99.2	—
3.5	101	99.9	98.4	97.8
3.0	100.9	99.8	99.5	98.3
2.0	100	99.5	99	98.9
1.0	99.4	99.8	99.6	99.7
0.0	100	100	100	100
−1.0	100.7	100.1	99.4	99.7
−2.0	101.5	101.1	99.5	99.8
−3.0	102.3	101.4	100.2	99.6
−3.5	102.2	101.2	99.7	98.6
−3.8	101.5	100.3	99.8	—
−4.0	100.5	99.1	99.1	97.1
−4.2	98.4	96.5	—	—
−4.3	—	—	97.2	95.3
−4.4	93.5	91.5	—	—
−4.6	83.5	81.8	92.5	91.6
−4.8	63.2	65.4	88.3	86.5
−5.0	43.5	47.7	75.7	77.7
−5.2	26.3	29.5	58.1	64.7
−5.4	15.8	18.1	45.3	49.7
−5.6	9.3	11.6	31.1	35.7
−5.8	7.5	8.4	22.2	24.9
−6.0	5.6	6.7	15.6	18
−6.2	5.6	5.9	12.8	14
−6.4	5.1	5.4	10.9	11.7
−6.6	4.9	5.1	9.8	10.4
−6.8	4.6	4.8	9.1	9.5
−7.0	—	—	7.9	8.8

It is observed that the agreement between the film and the ion-chamber measurements was better than ±2% in terms of absolute difference in the percent dose (dose relative to central axis) in the main beam. In the penumbra region the differences corresponded to positional error of up to 1 mm. The good agreement for open fields between the film and the ion-chamber measurements meant that the films could be used for DW profile measurements to confirm the FBX measurements since the jaw movement is the only mechanism involved to create the wedge distributions in an open field and hence a DW field has the spectra at depths comparable to those in the corresponding open fields. Therefore the film calibration factors obtained from the open fields could be used for the DW fields. The difference in processing condition was in terms of change in processing temperature (a change of about 3°C or more) and the strength of the processing chemicals.

Table 3.3 Off axis ratio (OAR) values obtained with tubes (PTs) array at three depths

	OAR D = 1.5 cm		OAR D = 10 cm	
Off axis position (cm)	**Film**	**PTs**	**Film**	**PTs**
9.9	—	—	2.9	2.4
8.5	2.9	6.2	3.2	4.1
7.1	4.8	9.0	10.1	12.3
5.7	12.5	17.3	39.3	47.5
4.3	115.9	108.1	115.2	110.1
2.8	113.7	112.5	110.4	109.5
1.4	106.4	105.3	105.3	106.3
0.0	100.0	100.0	100.0	100.0
−1.4	94.3	93.1	94.6	95.7
−2.8	89.2	90.3	88.9	87.5
−4.3	79.0	73.8	80.7	76.2
−5.7	10.0	15.3	25.3	34.2
−7.1	4.2	7.8	7.5	9.2
−8.5	2.6	6.5	2.9	4.6
−9.9	—	00	2.2	3.1

Note: The OARs are normalized at the centralized axis.

The OARs measured with PTs, NCTs and IVTs for the 45° DW at 1.5, 10 and 20 cm depths are given in Tables 3.3–3.5, respectively, along with corresponding film OARs.

It is observed that in the central region, the film and the FBX values agree closely, irrespective of the irradiation tube size, whereas towards the beam edges the differences among various tube sizes increase. As shown in Table 3.3, the OARs for the FBX (PTs) and the film agree to within ±1% (absolute difference in percentage dose calculated with respect to D_{max} value) in the central region at all depths whereas in the steep dose gradient region this difference corresponds normally to an average positional error of about ±2 mm with a maximum of about 2.5 mm. Table 3.4 shows that for the NCTs the agreement is within ±1.5% in the main beam region and within ±1.5 mm in the penumbra region. For the IVTs, Table 3.5 shows that the error in the central region is within ±1.9% and in the penumbra region it is about ±1 mm.

Table 3.4 Off axis ratio (OAR) values obtained with film and needle cover tubes

	OAR D = 1.5 cm		OAR D = 10 cm		OAR D = 20 cm	
Off axis position (cm)	**Film**	**NCT**	**Film**	**NCT**	**Film**	**NCT**
7.5	3.9	2.6	8.1	9.6	—	—
7.4	—	—	—	—	16.2	18.6
6.8	5.4	4.8	11.9	13	22.6	25.7
6.2	7.3	8.8	18.7	20.5	44.9	40.1
5.5	17.7	18.3	55.2	63.9	97.9	96.3
4.9	68.8	55.3	103	101.9	111.9	110.7
4.2	118.7	117.5	115.3	113.7	112.1	110.6
3.6	118	119.2	111.7	110.2	110.5	111.2
3.0	114.6	116.0	111.0	110.6	108.8	109.7
2.0	108.9	107.8	107.3	106.8	106.1	105.1
1.0	104.6	105.2	103.9	103.7	103.3	102.6
0.0	100.0	100.0	100.0	100.0	100.0	100.0
−1.0	95.7	94.2	96.1	95.3	96.3	95.4
−2.0	92.4	93.8	92.4	93.7	92.7	93.9

(Contiued)

Table 3.4 (*Continued*)

	OAR D = 1.5 cm		OAR D = 10 cm		OAR D = 20 cm	
Off axis position (cm)	Film	NCT	Film	NCT	Film	NCT
–3.0	88.6	90.1	88.2	87.1	88.8	89.6
–3.6	—	—	85.1	83.5	85.9	84.3
–3.7	85.4	84.5	—	—	—	—
–4.3	79.0	77.8	—	—	82.1	80.6
–4.4	—	—	79.2	78.2	—	—
–4.9	48.5	56.8	—	—	76.6	74.8
–5.0	—	—	63.7	65.9	—	—
–5.5	13.6	14.9	—	—	59.5	62.9
–5.6	—	—	28.5	33.8	—	—
–6.1	6.7	7.7	—	—	—	—
–6.2	—	—	13.0	11.4	24.3	27.9
–6.7	3.9	2.2	—	—	—	—
–6.8	—	—	9.1	10.2	15	18.2
–7.4	2.9	1.9	6.1	7.7	—	—
–7.5	—	—	—	—	11.9	9.5

Note: The OARs are normalized at the centralized axis.

Table 3.5 Off axis ratio (OAR) values obtained with film (IVTs) cathe tubes array at three depths

	OAR D = 1.5 cm		OAR D = 10 cm		OAR D = 20 cm	
Off axis position (cm)	Film	IVT OAR	Film	IVTs	Film	IVTs
69.	5.6	3.8	11.3	13.9	—	—
6.4	6.7	8.5	14.5	16.3	20.5	23.8
6.0	8.3	10.2	22.2	25.1	34.3	38.2
5.6	14.1	17.3	44.8	52.7	61.6	55.7
5.2	32.8	39.9	95.6	92.3	92.9	90.4
4.7	91.3	83.1	106.4	104.3	108.5	106.8
4.3	115.9	114.9	115.1	113.8	112.2	110.3

	OAR D = 1.5 cm		OAR D = 10 cm		OAR D = 20 cm	
Off axis position (cm)	**Film**	**IVT OAR**	**Film**	**IVTs**	**Film**	**IVTs**
3.9	116.8	118.2	114.2	112.4	111.3	111.9
3.4	119.7	117.9	112.6	113.7	109.9	11.3
3.0	114.6	112.6	111.0	112.9	108.8	110.4
2.0	108.9	107.2	107.3	105.8	106.1	105.7
1.0	104.6	105.9	103.9	104.5	103.3	104.2
0.0	100.0	100.0	100.0	100.0	100.0	100.0
−1.0	95.7	94.3	96.1	97.5	86.3	95.1
−2.0	92.4	93.2	92.4	93.9	92.7	94.3
−3.0	88.6	87.1	88.2	86.6	88.8	87.2
−3.4	87.1	88.5	86.0	87.2	86.4	85.2
−3.9	84.1	85.9	83.3	84.5	84.5	82.7
−4.3	79.0	77.5	80.4	78.1	82.1	81.5
−4.7	62.3	57.6	71.8	69.6	79.0	78.2
−5.2	26.6	31.7	53.1	57.9	71.2	68.2
−5.6	11.4	13.2	28.5	33.2	55.0	59.8
−6.0	6.5	8.1	15.0	17.3	32.4	36.5
−6.4	5.5	4.3	10.7	12.5	19.8	22.3
−6.9	4.7	2.7	8.8	9.2	14.1	11.5

Note: The OARs are normalized at the central axis.

The close matching of the FBX values with the film values in the central region irrespective of the tube size could be attributed to the fact that in this region the dose gradient is gradual and almost linear and hence the size of the tubes is not important. The small difference that is still observed may be due to the errors introduced mainly in the FBX measurements during various steps in the dosimetry procedure such as filling of tubes, placements of the same in the phantom and their subsequent read out in the colorimeter or the spectrophotometer. The overall random error is maximum (1.7%) in case of IVTs. This could be attributed to larger errors introduced during filling and extracting of a considerably small quantity (about 200 μL) of FBX solution from the sealed IVTs and transferring the same for reading in a 50 μL cuvette. The overall error values for PTs and NCTs are 0.7% and 1.1%, respectively. The larger difference between the IVT and the film measurements (1.9%) in the central (main beam) region as compared to that of PTs and NCTs could be due to the larger random error in the IVT case. The increased difference between the film and the FBX measurements in the steep dose gradient (penumbra) regions is mainly related to the size of the tubes; the larger tubes having lower spatial resolution. However, the improvement in spatial resolution in the beam edge regions for the IVTs as compared to NCTs is not in proportion to the reduction in tube size as is the case when going from PTs to NCTs. The reason again could be the larger random error associated with the IVTs relative to PTs and NCTs, offsetting the advantage associated with smaller tube size.

The dose delivered to the samples ranged from 9.8 to 170.2 cGy. When we compare data of Co-60 gamma rays, which clearly shows that the sensitivity of the FBX system for carbon beam is low as compared to that for gamma rays. At 100 cGy dose the optical density values are 0.028 ± 4.2% and 1.10 ± 2.2% for carbon and gamma rays, respectively. By comparing these values, we find the sensitivity ratio to be 25.5%. The overall uncertainty in dose estimation introduced due to individual uncertainties in various terms namely number of incident carbon particles (2.0%), carbon beam energy (0.003%), sample mass (0.7%), absorbance (4.2% and 2.2% for carbon beam and gamma rays, respectively) is 4.7% and 3.0% for carbon beam and gamma rays, respectively. In terms of random and non-random components, the uncertainty values are 4.2% and 2.1% for carbon and 2.2% and 2% for gamma rays, respectively. The overall uncertainty is calculated by summing the individual uncertainties in a quadratic manner. By this method, the uncertainty in the sensitivity ratio comes out to be 5.6%.

We also performed G (Fe^{3+}) calculations for comparison with published values. The formula used for calculation is based on Beer–Lambert law and can be written as

$$G(\text{molecules}/100\text{eV}) = \frac{(9.647 \times 10^{8} \times \Delta\text{OD})}{(D \times d \times \Delta e \times t)},$$

where D is dose (cGy), d is density (gm/cc) of the absorbing medium, OD is measured optical density or absorbance, t is path length (cm) in the cuvette, and e is molar absorption coefficient of ferric xylenol complex in M^{-1} cm^{-1}. The value of molar absorption coefficient (19,800 M^{-1} cm^{-1} ±0.5%) was taken from literature. The value of G (Fe^{3+}) thus obtained in molecules/100 eV was converted into new unit mol J^{-1}, conversion factor being 1 molecule (100 eV)$^{-1}$ equal to 1.036 × 10^{-7} mol J^{-1}. The G-values were found to be 14.2 × 10^{-7} mol J^{-1} (±4.7%) and 55.7 × 10^{-7} mol J^{-1} (±3.0%) for carbon beam and cobalt-60 gamma rays, respectively. The published G-value for gamma rays is 55.9 × 10^{-7} mol J^{-1}. Using the molar absorption coefficient value from literature as mentioned above instead of the one experimentally determined for our setup does introduce error in calculating the G(Fe^{3+}) value but the error will be marginal since the xylenol orange used by us is of the same make (Loba Chemie) as used by the authors quoted above. The G-value ratio of carbon and gamma-rays for FBX is 25.5% (±5.6%). In terms of random and non-random components, the uncertainties in the G-value estimation are 4.3% and 2.1% for carbon and 2.2% and 2.1% for gamma rays, respectively. It may be noted that the additional uncertainty component in the calculation of G-value as compared to the dose estimation done in the above paragraph, is due to the molar absorption coefficient (0.5%). We wish to point out that the determination of G-values and then the comparison of the G-value thus determined with the published G-value in the case of gamma rays was done in order to validate the experimental techniques adopted by us.

3.4 Summary and Conclusions

The literature survey on the applications of FBX dosimeter in radiotherapy showed that the FBX system had proved its utility in several types of dosimetric work with photons and electron beams. It was observed that in the reported work, the absorption measurements had been carried out using spectrophotometers and also the measurements were mostly associated with in-phantom studies under ideal laboratory conditions rather than a hospital environment. Laboratory work involved with the FBX system such as precision weighing of chemicals and preparing of solution require utmost care. Being a non-routine hospital activity, chances of compromise in precision and reproducibility of the dosimeter response therefore remain high while working in a hospital environment. Carrying out in vivo measurements brings the involvement of patients and with it certain other factors that may affect the dosimeter response. For example the irradiation tubes may either remain filled for long duration before irradiation or after irradiation or thus affect the reproducibility.

The work reported in this chapter, therefore, involved standardization of the FBX dosimetry system in hospital conditions utilizing the "in-house" facilities such as a colorimeter in place of a costly spectrophotometer. The reproducibility of the dosimeter was found to be within ±2% at 100 cGy and the same was considered satisfactory.

In the next part entrance dose measurements were carried out on patients undergoing external beam radiotherapy on a telecobalt machine. A total of 220 setups on 100 patients were measured for this study. The setups were simple parallel opposed fields; cranial irradiation (N = 100) and pelvic irradiation (N = 120). The reproducibility of the FBX dosimeter was also determined under clinical conditions for entrance dose measurement and was found to be better than ±4%. The results of the in vivo (entrance) dose measurements were compared with the published work on entrance dose measurements with diode detectors. The FBX performance was considered satisfactory from the comparison point of view and also from the role in vivo dosimetry plays in routine QA in radiotherapy.

The next study involved finding out the potential of FBX dosimeter in one of the current beam delivery techniques that is dynamic wedge. Measurement of dynamic wedge profiles was chosen as a typical case of radiation intensity variation, temporally and spatially. The dynamic wedge is a precursor to dynamic conformal therapy and therefore the dosimetry challenges are similar to the latter to some extent. Conventionally, it requires an elaborate setup such as a properly calibrated array of detectors—either ion-chambers or diode detectors—called LDAs or a film-based dosimetry system. The FBX dosimeter arrays were prepared from material available "in-house" such as needle covers and IV tube catheters and absorption measurements were performed on a colorimeter and a spectrophotometer. For comparison of the performance, profiles were measured with film. It was shown that the variation of the measurements was unacceptable for larger diameter FBX dosimeter tubes at the edges of the beams (equivalent positional errors more than 2 mm). However, for smaller tubes such as the ones made from IV catheters, the variations were within ±1 mm and thus satisfactory. It was noted that the time consumed in FBX dosimetry, at present, was high as compared to the commercially available methods. Improvisations were needed to make the FBX dosimetry system efficient before it could gain acceptability for routine measurements. Improvements were also needed to increase the spatial resolution. However, it was felt that the rather inexpensive FBX dosimeter could be used for routine QA of dynamic wedge dosimetry, even in the present form, where the costlier alternatives cannot be afforded. It was expected that without improvisations also, time spent can be reduced, to a great extent, with practice.

Protons and other heavy particles are increasingly been used in radiotherapy, the optimum mass being that of carbon or oxygen nuclei. The response of the aqueous FBX dosimeter was determined for a 55 MeV carbon to assess its potential in heavy-particle radiotherapy. Technique was developed for irradiation of the FBX dosimeter from a horizontally emerging low flux (10^3–10^4 cm^{-2} s^{-1}) beam. Small sachets were prepared from 6 μ thick polypropylene sheet and 200 μL solution filled in these sachets for irradiation. The response of the FBX system to a 55 MeV carbon beam was found to be 25.5% of gamma-ray response. Since for radiotherapy, the energy of the carbon beam or of any other charge particle will be significantly higher, the response of the FBX system will be higher for those beams. From the experimental studies described in this chapter, the following can be concluded:

(i) The FBX dosimetry system can be used for in vivo (entrance) dose measurements. The accuracy achievable with the system in external beam radiotherapy on a telecobalt machine is better than ±4%. This accuracy is achievable using a simple colorimeter for OD measurements.

(ii) The system can measure dynamic wedge profiles with the accuracy desirable in routine QA checks. However, in the present form, the procedure involved is more time consuming than the conventional LDAs and film-based system.

(iii) The FBX system holds potential for dosimetry in heavy-particle radiotherapy.

(iv) The FBX system is a simple and cost effective dosimetry system.

3.5 Future Scope

In order to validate new radiation treatment planning and delivery techniques aimed at 3D conformal RT, need of a 3D integrating dosimeter is felt. Gel dosimeters are apparently suitable for the purpose. However, Gel dosimeters have still not found a regular place among the various dosimetry systems routinely used in radiotherapy. The two possible reasons are limited access of clinical diagnostic MRI or CT in a hospital setup for dosimetry read-out and the perception that gel dosimetry is a complex procedure. With improved and streamlined procedures, and availability of better optical CT systems, Gel dosimetry is witnessing resurgence. The development of optical cone beam CT (OCBCT) has made the gel read-out affordable and faster removing the two major hurdles in clinical acceptability of gel dosimetry systems [28, 29]. At the same time, OCBCT has thrown up other kind of technical challenges such as errors due to scatter light and decrease in resolution restricting its applicability for radiochromic gels. There are reports in literature on improving the performance of the FXG system in terms of reducing scatter and diffusion in the system [20, 30], but more needs to be done. The future research should focus on ways to overcome some of the shortcomings of the radiochromic gel dosimetry systems.

References

1. Adyemi A, Lord J. An audit of radiotherapy patient doses measured with in-vivo semiconductor detectors. *Br J Radiol*, 1997; 70: 399–408.
2. Benedetto AR, Harris CR. Measurement of absorbed dose from radionuclide solutions mixed intimately with the FBX dosimeter. *Health Phys*, 1985; 49: 455–477.
3. Carver A, Gilmore M, Riley S, Uzan J, Mayles P. An analytical approach to acceptance criteria for quality assurance of intensity modulated radiotherapy. *Radiother Oncol*, 2011; 100: 453–455.

4. Chougule A, Hussain S, Gupta AR, Shrimali R. Use of FBX dosimeter for bladder dosimetry in patients undergoing intracavitary applications. *J Med Phys*, 1993; 18: 34–36.
5. Edwards CR, Grieveson MH, Mountford PJ, Rolfe P. A survey of current in-vivo radiotherapy dosimetry practice. *Br J Radiol*, 1997; 70: 299–302.
6. Feist H. Determination of the absorbed dose to water for high-energy photons and electrons by total absorption of electrons in ferrous sulphate solution. *Phys Med Biol*, 1982; 27: 1435.
7. Fricke H, Hart EJ. Chemical Dosimetry. In: (Attix FH, Roesch WC, eds), *Radiation Dosimetry*, vol II. 1966; New York: Academic Press.
8. Gupta BL. A low level chemical dosimetry system. *Proceedings of the Chemistry Symposium at IIT Madras*. 1970; vol II: pp. 49–55.
9. Gupta BL, Madhvanath U. Dosimetry of high energy photons and electrons using chemical dosimeters. *AMPI Med Phys Bull*, 1979; 1: 167–171.
10. Gupta BL. Low level dosimetry studies with the $FeSO_4$-benzoic acid-xylenol orange system. *IAEA International Symposium on Dosimetry in Agriculture, Industry, Biology and Medicine* (IAEA, Vienna) 1973; pp. 421–432.
11. Gupta BL. Current Status of FBX Dosimeter. Presented at 15th AMPI Annual Conference on Medical Physics. *Synopsis J Med Phys*, 1994; 19: 23.
12. Gupta BL, Bhat RM, Narayan GR. Stable FBX system for postal dosimetry of teletherapy machines. *AMPI Med Phys Bul*, 1987; 12: 6–8.
13. Gupta BL, Bhat RM, Gomathy KR, Suseela B. Radiation chemistry of the ferrous sulphate-benzoic acid-xylenol orange dosimeter. *Radiat Res*, 1978; 75: 269–277.
14. Gupta BL, Gomathy KR. Consistency of ferrous sulphate-benzoic acid-xylenol orange dosimeter. *Int J Appl Radiat Isot*, 1974; 25: 509–513.
15. Gupta BL, Kini UR, Bhat RM, Madhvanath U. Use of the FBX dosimeter for the calibration of cobalt-60 and high energy teletherapy machines. *Phys Med Biol*, 1982; 27: 235–245.
16. Gupta BL, Kini UR, Bhat RM, et al. Intercomparison of ionization and chemical dosimetry techniques for central axis depth-dose measurements from cobalt-60 teletherapy machines. *AMPI Med Phys Bull*, 1986; 11: 28–31.

17. Gupta BL, Nilekani S R. Scatter effects in electron dosimetry. *Int J Appl Radiat Isot*, 1980; 31: 41–43.
18. International Atomic Energy Agency. Absorbed dose determination in external beam radiotherapy. An International Code of Practice for Dosimetry. Technical Report Series 398, IAEA, Vienna (2000).
19. International Commission on Radiation Units and Measurements. Radiation Dosimetry: X Rays and Gamma Rays with Maximum Photon Energies Between 0.6 and 50 MeV.1969; ICRU Report No. 14: Washington DC.
20. Jordan K, Sekimoto M. Effects of adding glycerol and sucrose to ferrous xylenol orange hydrogel. *J Phys Conf Ser*, 2010; 250: 012048.
21. Kron T, Schneider M, Murray A, Mameghan H. Clinical thermoluminescence dosimetry: How do expectations and results compare? *Radiother Oncol*, 1993; 26: 151–161.
22. Lee PC, Sawicka JM, Glasgow GP. Patient dosimetry quality assurance program with a commercial diode system. *Int J Radiat Oncol Biol Phys*, 1994; 29: 1175–1182.
23. Leunens G, Van Dam J, Dutreix A, Schueren van der E. Quality assurance in radiotherapy by in-vivo dosimetry. 1. Entrance dose measurements, a reliable procedure. *Radiother Oncol*, 1990a; 17: 141–151.
24. Leunens G, Van Dam J, Dutreix A, Schueren van der E. Quality assurance in radiotherapy by in-vivo dosimetry. 2. Determination of the target absorbed dose. *Radiother Oncol*, 1990b; 19: 73–87.
25. Madhvanath U, Kini UR, Gupta BL. Calibration of brachytherapy sources using ferrous sulphate-benzoic acid-xylenol orange dosimeter. *Int J Appl Radiat Isot*, 1976; 27: 443–446.
26. Marshal TO. Accuracy and precision in thermoluminescence dosimetry. In: (Hufton AP, ed), *Practical Aspects of Thermoluminescence Dosimetry*. 1984; London: Hospital Physicists' Association.
27. Millwater CJ, Macleod AS, Thwaites DI. In-vivo semiconductor dosimetry as part of routine quality assurance. *Br J Radiol*, 1998; 71: 661–668.
28. Olding T, Darko J, Schreiner LJ. Effective management of FXG gel dosimetry. *J Phys Conf Ser*, 2010; 250: 012028.
29. Olding T, Holmes O, Schreiner LJ. Cone beam optical computed tomography for gel dosimetry I: scanner characterization. *Phys Med Biol*, 2010; 55: 2819–2840.

30. Penev KI, Mequanint K. Controlling sensitivity and stability of ferrous-xylenol orange-gelatin 3D gel dosimeters by doping with phenanthroline-type ligands and glyoxal. *Phys Med Biol*, 2013; 58: 1823–1838.

31. Pejuan A, Kuhn H. Properties of a modified Fricke solution for radiation dosimetry. *Proceedings of the Fourth Symposium on Neutron Dosimetry*, Munich—Neuherberg Commission of European Communities. *Radiat Prot*, 1981: 2: 455–464.

32. Ramani R, Russel CMD, O'Brien P. Clinical dosimeter using MOSFETs. *Int J Appl Radiat Isot J Radiat Oncol Biol Phys*, 1997; 37: 959–964.

33. Rikner G, Grussel E. General specifications for silicon semiconductors for use in radiation dosimetry. *Phys Med Biol*, 1987; 32: 1109–1117.

34. Semwal MK, Bansal AK, Thakur PK, Vidyasagar PB. FBX aqueous chemical dosimeter for measurement of virtual wedge profiles. *J Appl Clin Med Phys*, 2008; 9(4): 206–210.

35. Semwal MK, Bansal AK, Thakur PK, Vidyasagar PB. In-vivo (entrance) dose measurements in external beam radiotherapy with aqueous FBX dosimetry system. *J Med Phys*, 2005; 30: 32–35.

36. Semwal MK, Banerjee Milan, Sarma A, Vidyasagar PB. Response of the FBX system to a carbon beam: Its potential as a dosimeter in heavy particle radiotherapy. *Phys Med Biol*, 2002; 47: N179–N183.

37. Shortt KR, Klassen NV, Ross CK, et al. Ferrous sulphate dosimetry and its role in establishing an absorbed dose to water standard for the National research Council of Canada. Workshop on Water Calorimetry. Rep. NRC-29637 (Ross CK, Klassen NV eds), National Research Council of Canada, Ottawa. 1988: 121–126.

38. Chung WH. Chemical dosimetry. In: (Mahesh K, Vij DR, eds), *Techniques of Radiation Dosimetry*. 1985; New Delhi: Wiley Eastern Publishers.

Chapter 4

Radiobiological Effects in Fractionated Radiotherapy of Head and Neck Squamous Cell Carcinoma Patients

Arumugham Balraj,[a,b] Pandit B. Vidyasagar,[a] N. Chakravarty,[b] P. K. Thakur,[b] and S. Bhatnagar[b]

[a]*Biophysics Laboratory, Department of Physics, University of Pune, Pune, India*
[b]*Department of Radiation Oncology, Malignant Disease Treatment Centre, Command Hospital (SC), Pune, India*

balumano@gmail.com

Radiobiological effects depend greatly on the nature of the tissue in question and how the physical dose is received which is defined by the treatment time (T), dose rate (D), and fractionation scheme (F). Time dose fractionation (TDF) determines the radio biological effective dose (BED). The BED should be in the optimum range for better disease-free survival (DFS). In this study, 100 patients were treated with conventional fractionation regime in arm 1 and equal number of patients treated with the hyperfractionation regime in arm 2. With the patient's regular follow-up, the 5-year DFS were studied by analyzing the BED values of each patient. In arm 1, the patients were treated with parallel opposing fields in the head & neck region. In each field 90 or 100 cGy

Radiation in Medicine and Biology
Edited by Pandit B. Vidyasagar, Sagar S. Jagtap, and Omprakash Yemul
Copyright © 2017 Pan Stanford Publishing Pte. Ltd.
ISBN 978-981-4745-92-5 (Hardcover), 978-1-315-20656-1 (eBook)
www.panstanford.com

doses were given so that 180 or 200 cGy doses were delivered at the center of the tumor. All patients were treated 5 days per week for 6 weeks continuously. In arm 2, the doses delivered were 110 cGy in the morning and after a gap of 6 h again the same scheduled is followed. The treatment went on 5 days per week and a total of 33 treatment days. Chemotherapy agent cisplatin was administered to these patients once weekly to sensitize the tumor to radiation therapy. In both the regimes, BED was calculated for tumor (BEDta), normal cell early reactions (BEDne), and normal cell late reactions (BEDnl). The patients 5-year DFS were studied for both the regimes against the BEDta, BEDne and BEDnl. Five-year DFS in arm 1 was 9% whereas in arm 2 it was 24%. In both the regimes, 33% patients had more than 5-year DFS. The optimum BED can give better DFS, which can be achieved by the hyperfractionation regime.

4.1 Introduction

Radiotherapy is often the primary treatment of locally advanced squamous cell head and neck cancer, but the optimal fractionation schedule has been controversial [1]. The aim of clinical radiotherapy is to obtain for each cancer patient the highest probability of tumor control while minimizing the probability of late complications [4]. In radiotherapy it is the most difficult task of balancing between tumor control and late complications to arrive at the most appropriate external beam therapy fractionation schedule, dose rate per fraction, overall treatment time, and total dose [13]. The distribution of radiation dose over the time, known as fractionation, is one of the most important factors determining the outcome of the radiotherapy [11]. Since Regaud established that fractionation of the radiation doses improved the results of the radiotherapy as compared to the single dose. Accordingly, radiation schedules consisting of dose per fractionation of about 200 cGy daily, over several weeks have become conventional in clinical practice (6). However, the conventional fractionation schedules have resulted in poor tumor control rates in advanced malignancies [16]. The search for optimal fractionation schedules which can produce better tumor control rate in advanced malignancies has given rise to various fractionation schedules [1]. For classifying the various modified fractionation schedules two different approaches have been proposed [18]. In one, the objective is to maintain the dose per fraction as in conventional radiation schedule (about 2 Gy) and two or more than two fractions per day are delivered. In this schedule the total dose is administered in shorter overall time and the name proposed for this type of schedule is that of Accelerated fractionation or "rapid fractionation" [28]. The second approach is to administer a lower dose per fraction (75–150 cGy) several times per day. The total dose will be delivered by increasing the number of fractions and because of this these radiation schedules are known as "hyperfractionation" [28].

A variety of fractionation schedules including standard fractionation, hyperfractionation, accelerated fractionation, and their variants have been used in the radiotherapy of the advanced head and neck cancer [2, 3, 27]. In essence, altered fractionation is predicted to improve the therapeutic ratio through a differential response between tumors and normal tissues to fractionated radiotherapy. Theoretically, with hyperfractionation it is possible to increase the total dose, thereby increasing the probability of tumor control without increasing late complications through the use of multiple smaller dose fractionations [1, 11].

An important component of comparing the efficacy of various fractionation and dose rate regimes is their effect on tumor control and normal tissues. The expectations for a therapeutic gain from the implementation of these treatment schedules are mainly based on radiobiological modeling [24]. Dose fractionation and dose rate in radiotherapy has evolved along divergent paths in different centers worldwide [8]. It has been observed that in our Malignant Disease Treatment Centre, the head and neck cancer patients could not tolerate the different fractionation scheduled followed by the Western countries, due to low hemoglobin level, unhygienic conditions, and poor socioeconomic status. In view of these, we have tried all the fractionation schedules followed by the other centers worldwide and established the hyperfractionation schedule suitable to our patients. The BED was calculated for different patients and the 5-year DFS has been analyzed for conventional fractionation and hyperfractionation schedule.

4.2 Materials and Methods

Patients 18 years or older with Karnofsky Performance Score (KPS) (*Oxford Textbook of Palliative Medicine*, Oxford University Press, 1993) were selected. The KPS Index allows patients to be classified according to their functional impairment. This can be used to compare effectiveness of different therapies and to assess the prognosis in individual patients. The lower the Karnofsky score, the worse the survival for most serious illnesses. Patient with KPS ≥ 60 who had previously untreated stage III or IV but M_0 (non-metastasis) squamous cell carcinoma of the oral cavity, oropharynx, supraglottic larynx, base of tongue, hypopharynx, nasopharynx, or stages II to IV of carcinoma of larynx were included. Patients with a prior (within 5 years) or synchronous malignancy other than non-melanoma skin cancer were excluded.

Examination such as medical history, and physical examination, complete blood count, chest X-Ray, computed tomography or magnetic resonance imaging of the head and neck, a diagram of the primary tumor and neck nodes, and a dental evaluation were carried out. The disease was staged according to the 2001 American Joint Committee on Cancer (AJCC) Staging. All patients were treated as per the 2001 National Comprehensive Cancer Network (NCCN) clinical practice guidelines in oncology. A study specific informed consent was obtained from all patients. Eligible patients were randomized to receive radiotherapy delivered using (1) conventional fractionation at 1.8 or 2.0 Gy per fraction, one fraction per day and (2) hyperfractionation at 1.1 Gy per fraction, two fractions per day using cobalt-60 gamma rays as an external beam radiation treatment.

4.2.1 Conventional Fractionation

The patients those who were put on conventional fractionation method were treated with parallel opposing fields in the head and neck region viz., left lateral and right lateral. In each field, 90 or 100 cGy doses were given so that 180 or 200 cGy dose was delivered at the center of the tumor. All patients were treated 5 days per week and 6 weeks continuously; therefore the total dose delivered was around 54 to 60 Gy, depending upon the general health condition of the patient.

4.2.2 Hyperfractionation

The patients those who were put on the hyperfractionation method were treated with parallel opposing fields in the head and neck region same as the conventional fractionation patients. Here the difference was that the patients were treated in the morning as well as in the evening of the same day, that is, two times in a day. The doses delivered were 55 cGy for left lateral and 55 cGy for the right lateral in the morning and again after a gap of 6 h, the same schedule was followed. Therefore, the patients received a dose of 220 cGy per day. The treatment went on 5 days per week and total 33 treatment days, so that each patient was delivered dose of 72.6 Gy. [17]. For both fractionation regimes, the chemotherapy agents cisplatin or carboplatin of recommended dose (50 mg of cisplatin or 125 mg of carboplatin) were administered to all patients except the stage II of larynx patients, weekly once to sensitize the tumor to radiation therapy [1, 5].

4.2.3 External Beam Radiation Therapy on Theratron-780E Co-60 Teletherapy Machine

Radiotherapy was delivered using Theratron 780E Co-60 Teletherapy machine with source to surface or source to isocenter distance ≤ 80 cm. Before starting the actual external beam radiation therapy (EBRT), each patient had undergone simulation on Simulix HP 1200 simulator manufactured and supplied by Nucletron Holland. Beam interactions on anatomical region have been verified by simulation. For permanent records and for future reference and also for field reduction during the treatment period, the verification port films (X-ray films of the simulator) of the simulation have been obtained for each patient. A combination of lateral opposing fields, anterior and lateral wedged fields or other field arrangements were used to treat primary tumor and the lymph nodes in the upper neck. A single anterior field was used to treat the neck below the fields for the primary tumor. For patients with nodes more than 6 cm, supraclavicular nodes, or pyriform sinus tumors that were T3 or T4 or with clinically positive nodes, the anterior field could extend 5 cm inferiorly to include the upper mediastinum. All fields were treated on each treatment days. At least one field reduction was done for all patients in both the arms. The first field reduction, for the spinal cord off was done at 40–44 Gy and second field reduction done at 54–60 Gy for hyperfractionation patients. Whereas for conventional fractionation patients, only a single field reduction was implemented for the spinal cord off at 40 Gy. In the third field reduction the additional boost not exceeding 5–6 Gy through reduced fields to persistent primary tumor and or clinically positive nodes were allowed. A minimum 2 cm margin around the initial tumor volume and positive neck nodes for the first field reduction, and a minimum 1 cm margin for the third field reduction were implemented. The primary treatment fields were reduced off the spinal cord at 40–44 Gy. Neck dissection was allowed for neck nodes more than 3 cm in size (diameter) prior to radiotherapy at the discretion of the responsible head and neck surgeon, and radiation oncologist. Each case was discussed in detail in the weekly tumor board meetings and ENT clinical board meetings and the decision was made by the board about the kind of treatment schedule.

4.2.4 Follow-Up

During treatment, patients were clinically examined weekly, and blood tests were also done on a weekly basis. For all eligible conventional fractionation patients and hyperfractionation patients, chemotherapy agents cisplatin or carboplatin were administered every week, especially on Sundays. Once the treatment was ended, patients were evaluated within a month (4 weeks) and after 3 months for the second follow-up and after 3 months for the first 2 years and 6 months for the next 2–3 years and annually after 5th year onward. In addition to the tumor status, acute and late (occurring > 90 days from start of the treatment) normal tissue effects were graded with KPS (Table 4.1).

4.2.5 Quality Control and Quality Assurances

Radiotherapy records of each patients Simulation, Treatment planning, Treatment time calculation by accurately measuring the tumor depth, obtaining the simulation films, total dose, number of fractions, and elapsed treatment days relative to the protocol prescriptions were noted and taken care of for each patient. Since the dose delivery was done with the Theratron 780E Co–60 Teletherapy machine, the quality control and quality assurance tests of the Co–60 machine were carried out periodically as per the international standard protocol. Beam output measurements were carried out systematically by measuring the beam by calibrated secondary standard dosimeter (SSD) and the beam output was compared with International Atomic Energy Agency (IAEA) postal dose intercomparison conducted by Bhabha Atomic Research Centre (BARC), Mumbai.

4.3 Results and Discussion

4.3.1 Characteristics of the Patients

Between October 2001 and January 2007, 250 patients with squamous cell carcinoma of the head and the neck were treated in this study. Of these, approximately 5% of the patients discontinued the treatment in between due to their personal problems; 12% of the patients who had inadequate data were excluded from this study; 3% of the patients did not have adequate follow-up. The remaining patients numbered 200, who received the treatment as planned for this analysis. Out of these patients, 100 patients in the arm 1 (Group: 1) were treated with the conventional fractionation schedule and remaining 100 patients in arm 2 (Group: 2) were treated with the hyperfractionation schedule. Table 4.1 shows the pretreatment patient characteristics and group statistics in respect of both conventional (Group 1) and the hyperfractionation (Group 2) regime. They were well balanced among the treatment groups.

In Table 4.1 all the mean values of the characteristics are almost same for both the regimes; therefore the patients' treatment outcome is comparable in both regimes. However, the mean values in respect of dose related factors, viz., time, dose, fraction size, BEDta, BEDne, BEDnl, and DF (dose per fraction), are different, since these are the main differences between conventional and hyperfractionation regimes.

Table 4.1 Patients characteristics in both conventional (group 1) and hyper fraction (group 2) regime

	Group Statistics				
	Gr	N	Mean	Std. Deviation	Std. Error
Age	1	100	52.86	13.788	1.379
	2	100	56.69	11.628	1.163
IniKPS	1	100	75.30	5.766	.577
	2	100	74.30	5.904	.590
Con Dose	1	100	5851.80	547.788	54.779
	2	100	6975.80	633.204	63.320
Fraction size	1	100	30.69	3.051	.305
	2	100	63.22	6.301	.630
RT days	1	100	43.38	6.075	.607
	2	100	46.25	6.571	.657
BEDta	1	100	55.183	5.3073	.5307
	2	100	61.826	5.8633	.5863
BEDne	1	100	60.817	5.4618	.5462
	2	100	67.509	5.8165	.5816
BEDnl	1	100	84.577	8.4708	.8471
	2	100	89.429	8.0777	.8078
DF	1	100	1.911	.1205	.0120
	2	100	1.105	.0429	.0043
Weight	1	100	55.04	7.959	.796
	2	100	58.53	8.911	.891
Height	1	100	159.11	6.477	.648
	2	100	163.55	5.904	.590
BMI	1	100	21.7256	2.69685	.26969
	2	100	21.9296	3.41137	.34114
Fd size	1	100	180.04	65.604	6.560
	2	100	192.14	71.157	7.116
Hbstart	1	100	12.119	11.1225	1.1123
	2	100	11.368	1.6557	.1656
Wtloss	1	100	2.96	.650	.065
	2	100	2.95	.968	.097
Wtgain	1	100	2.32	.723	.072
	2	99	2.86	.742	.075

In Table 4.2, in the overall case distribution the oral cancer and oropharyngeal cancer accounts more in number that is 47 and 87, respectively. Most of the international multi-institution joint studies also show that the occurrence of oral and oropharynx cancer is more than the rest of the cancer types in the head and neck region [11]. The study populations (case distribution) are not homogeneous with respect to diagnostic classification ($p < 0.01$, the value of χ^2 is 15.96). However, this would not affect the duration of diseases-free survival (DFS), which is the outcome of the study. The staging of the disease is done as per the National Comprehensive Cancer Network (NCCN) guidelines. The staging depends upon the tumor size (T) and node sites and number of nodes as well as size of the nodes (N) and also the spread of the disease itself, which is described as metastasis (mets) (M). Therefore, the final staging depends upon TNM classifications (NCCN guidelines on cancer practice, 2007). The staging distributions are also given in Table 4.2.

Table 4.2 Case distribution as per the anatomical sites and diseases stage of head and neck cancer in conventional fractionation (Group 1) and hyperfractionation (Group 2) regimes

Diagnosis/stage	Group 1	Group 2
Ca Larynx	11	21
Oral Cancer	36	11
Ca Oropharyngeal	29	58
Paranasal Sinus	19	04
Ca Naso Pharynx	05	06
Total	100	100
Stage II	17	12
Stage III	39	21
Stage IV	40	50
Stage IVB	04	17
Total	100	100

As per the disease stage distribution, stage III and stage IVA of the disease comprise around 75% in the overall distribution. The stage IV B patients are more in the hyperfractionation schedule. The surgically unresectable of stage IVA and IV B comprise as 44 in conventional fractionation and 67 in hyperfractionation. This clearly shows that the highly advanced surgically unresectable tumors are selected for hyperfractionated regime along with concurrent chemotherapy than the conventional fractionation along with concurrent chemotherapy. The stage distribution of the disease shows statistically significant difference (p = 0.001, χ^2 = 15.42).

4.3.2 Disease-Free Survival

In the field of oncology, the no evidence of the disease (NED) has been appropriately termed as DFS, since the patient might live without the disease for some time, and in most of the cases, the recurrence occurs only after the passage of some considerable time [29]. The duration between the treatment and the recurrence of the disease is termed as the DFS period. Therefore, DFS is one of the most important factors of the outcome of this study. Although loco regionally controlled (LRC) disease and overall survival (OAS) play the role of the final outcomes of the treatment, the DFS is the most important factor for the outcome of the treatment modality.

The statistical data analysis has been done for all the 200 patients and the SPSS statistical analysis computer software has been used for analyzing the data. The following graph 1 shows the difference of the DFS in the conventional fractionation regime and the hyperfractionation regime. In the survival curve of the graph 1, green represents the hyperfractionation regime and blue represents the conventional fractionation regime. The number of patients in disease-free survival state for more than 60 months, in the hyperfractionation regime was 24%, whereas 09% in conventional fractionation regime. Overall DFS for more than 60 months in both the regimes was 16.5%. These findings have been seen in many large randomized trials that had compared hyperfractionated RT with conventional fractionated RT [1, 7, 10–12, 14, 15, 19, 22].

As per the results of the above disease-free survival graph 1 (Fig. 4.1), it is clearly indicated that the DFS is better in the hyperfractionation regime than the conventional fractionation regime. The fractionation scheme, whether it is conventional or hyper or accelerated or modified fractionations, depends upon the time (T), dose (D), fraction (F), called generally TDF factor

[20, 21]. The BED is calculated with this TDF factors as well as characteristics of different tissues and particular tissue responses in the same tissue. The BED formulas are readily available from the literature for different type of cancers and its tissues [25]. Biological effective doses can be calculated for different fractionation scheme (BED for Hyperfractionation or BED for Conventional fractionation, etc.). Therefore, DFS depends upon the TDF factors. In other words, it depends upon BED. BED can be calculated for tumor (BEDta), normal cell early reactions (BEDne), and normal cell late reactions (BEDnl).

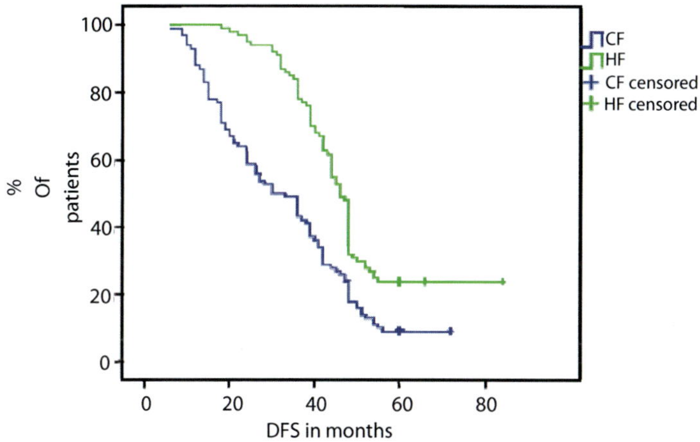

Figure 4.1 Disease-free survival (DFS) curve for conventional fractionation (CF) vs. hyperfractionation (HF) in head and neck cancer patients.

4.3.3 Formula for Calculation of BED for Conventional Fractionations

(1) BED for tumor (BEDta)
[(Total dose/100) × (1 + dose per fraction/13.4)] × [(0.76 × (TRT days-28)]
TRT days = Total radiotherapy days (starting to end)

(2) BED for normal cell early reactions (BED ne)
[(Total dose/100) × (1 + dose per fraction/11.2)] × [(0.50 × (TRT days-28)]

(3) BED for normal cell late reactions (BEDnl)
[(Total dose/100) × (1 + dose per fraction/4.3)]

4.3.4 Formula for Calculation of BED for Hyperfractionation

(1) BED for tumor (BEDta)
[(Total dose/100) × (1 + ((dose per fraction/13.4) × 1.031)] × [(0.76 × (TRT days-28))

(2) BED for normal cell early reactions (BEDne)
[(Total dose/100) × (1 + ((dose per fraction/11.2) × 1.000978)] × [(0.76 × (TRT days-28)]

(3) BED for normal cell late reactions (BEDnl)
[(Total dose/100) × (1 + (dose per fraction/4.3) × 1.099)]

4.3.5 Conventional BED Calculations

For example, a patient undergoes radiotherapy in conventional fractionation regime and received 6600 cGy in 33 fractions for 42 days. Here total dose is 6600 cGy, Dose per fraction is 6600/33 = 200, it is taken as 2 Gy, and total RT days is 42. In this case, the calculated BEDta = 65.2, BEDne = 70.8, BEDnl = 96.7.

4.3.6 Hyperfractionation BED Calculations

For example, a patient undergoes radiotherapy in the hyperfractionation regime and received 7260 cGy in 66 fractions on 46 days. Here total dose is 7260 cGy, dose per fraction is 7260/66 = 110, it is taken as 1.1 Gy, and total RT days is 46. In this case the calculated BEDta = 65.1, BEDne = 70.7, BEDnl = 93.0.

4.3.7 Comparison of Conventional and Hyperfractionation of BED Calculations

Table 4.3 Comparison of conventional and hyper fractionation of BED calculations.

Factors	Conventional	Hyper
Total dose, (cGy), total fraction, total RT days	6600/33 fraction/42 days	7260/66 fraction/46 days
BEDta	65.2	65.1
BEDne	70.8	70.7
BEDnl	96.7	93.0

From Table 4.3, it is understood that the conventional fractionation and hyperfractionation, though the BEDta and BEDne values are almost same the BEDnl value in the hyperfractionation regime is 93.0 compared to 96.7 in the conventional regime. This normal cell late reaction, which is lesser in hyperfractionation than the conventional fractionation, shows the DFS more in the hyperfractionation regime. This study suggests that total dose, total treatment duration (overall time), and the number of fractions are important in the outcome of the radiotherapy. The BED should be adequate and it should be in the optimum range for better DFS. Although the biological effective dose, play a major role for better disease-free survival, the optimum biological effective dose, which may have a better disease-free survival, may be achieved by the hyperfractionation regime. Using chemotherapy along with radiotherapy also improves the DFS. Several toxicity antagonists are under active investigation [26]. In the future, some of these agents may decrease the acute and late effects of cancer therapy [1]. The DFS may also be improved by using conformal and intensity modulated radiotherapy (IMRT) and image guided radiotherapy (IGRT). These modern radiotherapy techniques have the capability of improving tumor target coverage while minimizing the dose to and volume of the surrounding normal tissue irradiated [9]. Further studies are needed to determine whether these modern radiotherapy techniques improve the outcome of the fractionated radiotherapy and thereby decreasing the acute and late effects and improving the disease-free survival.

References

1. Azza Abd El-Naby et al. (2008) Conventional versus Hyperfractionated radiotherapy in locally advanced head and neck cancer. *Cancer Ther*, **6**: 673–682.
2. AnG KK. (1998) Altered fractionation trials in head and neck cancer. *Semin Radiat Oncol*, **8**: 230–236.
3. Ang KK, Peters LJ. (1994) Principles and practices of oncology, *PPO Updates*, **4**: 1–15.
4. Cox JD, Pajak TF, Marcil VA et al. (1990) Dose response for local control with hyperfractionated radiation therapy in advanced

carcinomas of the upper aero digestive tracts: Radiation therapy Oncology Group protocol 83–13. *Int J Radiat Biol Phys,* **18**: 515–521.

5. Cox JD, Stetz J, Pajak TF. (1995) Toxicity criteria of the radiation therapy oncology group (RTOG) and European Organisation for research and treatment of cancer (EORTC) (editorial). *Int J Radiat Biol Phys,* **31**: 1341–1346.
6. Coutard H. (1934) Principles of X-ray therapy of malignant disease. *Lancet,* **2**: 1–8.
7. Cummings BJ, Keane TJ, Pintilie M, O'Sullivan B, Payne D, Warde P, McLean M, Waldron J, Liu FF, Gullane P, ENT Group (1996) A prospective randomized trial of hyperfractionated versus conventional once daily radiation for advanced squamous cell carcinomas of the pharynx. *Int J Radiat Oncol Biol Phys,* **36**(Suppl 1): S235.
8. Dische SA, Saunders M, Barret A, Harvey A, Gibson D. Parmar MA. (1997) Randomized multi centre trial of CHART versus conventional radiotherapy in head and neck cancer. *Radiother Oncol,* **44**: 123–136.
9. Eisbruch A. Ten Haken RK, Kim HM, Marsh LH, Ship JA. (1999) Dose volume and function relationships in parotid salivary glands following conformal and intensity-modulated irradiation of head and neck cancer. *Int J Radiat Biol Phys,* **45**: 577–587.
10. El-Sayed S, Nelson N (1996) Adjuvant and adjunctive chemotherapy in the management of squamous cell carcinoma of the head and neck region: A meta-analysis of prospective and randomized trials. *J Clin Oncol,* **14**: 838–847.
11. Fu KK, Pajak TF, Trotti A, Jones CU, Spencer SA, Phillips TL, Garden AS, Ridge JA, Cooper JS, Ang KK (2000) A radiation therapy oncology group (RTOG) phase III randomized study to compare hyperfractionation and two variants of accelerated fractionation to standard fractionation radiotherapy for head and neck squamous cell carcinomas: First report of RTOG 9003. *Int J Radiat Oncol Biol Phys,* **48**: 7–16.
12. Galiana R, Boladeras A, Mesía R, Gómez J, de Juan A, Manos M, Nogués J, Navarro V, Guedea F (2002) Twice-a-day radiotherapy for head and neck cancer: The Catalan Institute of Oncology experience. *Radiother Oncol,* **64**: 19–27.
13. Harari PM, Fowler F. (1994) Idealised versus realized overall treatment times. *Int. J. Radiat Oncol Biol Phys,* **29**: 209–211.
14. Horiot JC, Le Fur R, N'Guyen T (1992) Hyperfraction versus conventional fraction in oropharyngeal carcinoma: Final analysis

of a randomized trial of the EORTC cooperative group of radiotherapy. *Radiother Oncol*, **25**: 231–241.

15. Jeremic B, Shibamoto Y, Milicic B, Nikolic N, Dagovic A, Aleksandrovic J, Vaskovic Z, Tadic L. (2000) Hyperfractionated radiotherapy with or without concurrent low-dose daily cisplatin in locally advanced squamous cell carcinoma of the head and neck: A prospective randomized trial. *J Clin Oncol*, **18**: 1455–1464.
16. Skladowski K, Maciejewski B, Golen M, Pilecki B, Przeorek W, Tarnawski R. (2000) Randomized clinical trial on 7-day continuous accelerated irradiation (CAIR) of head and neck cancer—report on 3 year tumour control and normal tissue toxicity. *Radiother Oncol*, **55**(2): 101–110.
17. Marcial VA, Pajak TF. (1987) Hyper fractionated photon radiation therapy in the treatment of advanced squamous cell carcinoma of the oral cavity, Preliminary report by the Radiation Therapy Oncology Group (RTOG). *Int J Radiat Oncol Biol Phys*, **13**: 41–47.
18. Poulsen MG1, Denham JW, Peters LJ, et al. (2001) Randomized trial of accelerated and conventional radiotherapy for stage III and IV squamous cell carcinoma of the head and neck a Trans–Tasman Radiation Oncology group study. *Radiother Oncol*, **60**: 113–122.
19. Orecchia R, Jereczek-Fossa BA, Catalano G, Chiesa F, De Pas T, Masci G, Krengli M, Vavassori A, De Paoli F, Robertson C, Marrocco E, De Braud F (2002) Phase II trial of vinorelbine, cisplatin and continuous infusion of 5-fluorouracil followed by hyperfractionated radiotherapy in locally advanced head and neck cancer. *Oncology*, **63**: 115–123.
20. Orton CG. (1992) Time dose models. In: *Handbook of Medical Physics* (Waggener RG, ed), vol I, CRC Press, pp. 265–268.
21. Orton CG, Cohen L. (1988) A unified approaches to dose-effect relationships in radiotherapy I: Modified TDF and linear quadratic equations. *Int J Radiat Oncol Biol Phys*, **14**: 549–556.
22. Pinto LH, Canary PC, Araújo CM, Bacelar SC, Souhami L (1991) Prospective randomized trial comparing hyperfractionated versus conventional radiotherapy in stages III and IV oropharyngeal carcinoma. *Int J Radiat Oncol Biol Phys*, **21**: 557–562.
23. Regaud C. (1922) Influence de la dur se irradiation sur les effect determines dons la testable par le radium. *Com. Rend. Soc. Biol.* **86**: 787–798.

24. Supe SJ. (1984) Time dose fractionation factors for tumours. In: *Optimization of Cancer Radiotherapy* (Paliwal BR, ed), AAPM Publication—Symposium Proceedings, No. 5, New York, pp. 339–345.
25. Supe SS. (1994) The Application of linear quadratic model of dose effect relationship to radiotherapy. Ph.D. Thesis, Marathwada University.
26. Trotti A. (2001) Toxicity antagonists in cancer therapy. *Curr Oncol,* **9**: 569–578.
27. Withers HD. (1995) Biologic basis for altered fractionation schemes. *Cancer,* **55**: 2086–2095.
28. Withers HR, Thames HD, Peters LJ. (1982) Differences in the fractionation response of acute and late-responding tissues. *Prog Radiat Oncol,* **II**: 257–296.
29. Perez CA, Brady LW. (2008) Principles and Practice of Radiation Oncology, 5th ed., Lippincott Williams & Wilkins.

Chapter 5

Radio-Electro-Chemotherapy of Cancer: New Perspectives for Cancer Treatment

Pratip Shil,[a] Pandit B. Vidyasagar,[b] and Kaushala Prasad Mishra[c]

[a]*Bioinformatics Laboratory, National Institute of Virology,*
130/1 Sus Road, Pashan, Pune, Maharashtra, India
[b]*Vice-Chancellor, Swami Ramanand Teerth Marathwada University,*
Nanded, Maharashtra, India
[c]*Ex-Vice Chancellor, Nehru Gram Bharti University,*
104 F/3, Malviya Road, George Town,
Allahabad, Uttar Pradesh, India

shilpratip@gmail.com

5.1 Introduction

The progress of biomedical sciences to the current level owes deeply to the constructive role played by physical and biophysical principles, techniques, and tools. This extends to all branches of medical sciences, including cancer research. Cancer is a dreaded disease that continues to kill millions of people worldwide. Considerable efforts have, therefore, been directed to research for better understanding of tumorigenesis, prevention and therapy. The major modalities of cancer treatment so far developed are surgery, chemotherapy, and radiotherapy. However, these

Radiation in Medicine and Biology
Edited by Pandit B. Vidyasagar, Sagar S. Jagtap, and Omprakash Yemul
Copyright © 2017 Pan Stanford Publishing Pte. Ltd.
ISBN 978-981-4745-92-5 (Hardcover), 978-1-315-20656-1 (eBook)
www.panstanford.com

have inherent limitations, and the major problem faced by clinicians and researchers is the development of radio-resistance and chemo-resistance by cancer cells. Also, the side effects of anticancer drugs severely affect the patients. Therefore, there is a need for the development of new approaches and optimization of protocols for overcoming these difficulties and provide effective cancer therapy to the patients. Thus, the development of effective therapy of cancer is one of the major challenges of the 21st century.

Physical and biophysical techniques have played an important role in the diagnostics and therapeutic strategies for the treatment of diseases. In this respect, use of biophysical principles, methodologies, and instruments has provided new and sensitive approaches to cancer treatment. However, there is a need to enhance the effectiveness of existing treatments of radiation and chemotherapy.

In order to increase the intracellular drug accumulation, various techniques have been developed such as incorporation of drugs into liposomes, using specific antibodies, magnetic drug targeting or enhancing the membrane permeability by physical or chemical means [1–6]. Against this background, it is remarkable to notice that externally applied electric pulses on cells and tissues have provided a new approach to achieving enhanced killing of cells in vitro and in vivo. This method is called electroporation. Electroporation is one of the few biophysical techniques that have grown rapidly from the laboratory to application in cancer therapy clinics [7–13].

In this chapter, we discuss the possibilities in enhancing the effectiveness of the radiation and chemotherapy treatments when combined with electroporation. Investigations have been made on animal systems (in vivo) and at the cellular level (in vitro).

5.2 Electroporation Technologies

Electroporation involves the enhancement of the plasma membrane permeability by exposing the biological cells to electric pulses of high voltage and short durations. As early as 1967, Sale and Hamilton [14] had observed the killing of microorganisms, viz., bacteria and yeast due to exposure to high-voltage electric pulses. It was observed in 1972 by Neumann and Rosenheck [15] that the exposure to electric field generated changes in the permeability of the biological cells. This was followed by the study of electric field induced pore formation in plasma membrane by Kinosita and Tsong [16].

Abidor et al. (1978) [17] investigated the effects of electric field on biological cells and gave a detailed physical theory about electroporation. He also proposed the mathematical explanation of pore formation. These pores were transient and vanished soon after the withdrawal of the electric field. The micro-pores allow direct access to the cytosol and molecules, which are otherwise impermeable, and can enter the cytosol. Since then electroporation emerged as a powerful technique for biomedical research with a wide range of applications. Along with the development of application-based technologies, theoretical models based on physical laws have been proposed for a better understanding of the phenomenon. These involve measurements of electrical behavior for cell and model planer membranes as a function of time [18–23] and number of molecules of different shapes, size, and charge that are taken up by the cells for various electric pulses pertaining to both average cellular uptake [24–26] and at the individual cellular level [27–30].

5.3 Biophysical Basis of Electroporation

The biological cell can be imagined to be a nonconducting sphere with an equipotential inner side. For such a cell, placed in an externally applied electric field E_e, the transmembrane potential at any point on the cell membrane is obtained from Laplace's Equation and is given by

$$\Phi_m = 1.5\, r_{cell} E_e \cos\theta, \tag{5.1}$$

where r_{cell} is the radius of the cell and θ the angle between the radius vector of the point at which the potential is measured and the direction of the electric field and the factor 1.5 represents a constant for the spherical geometry of the cell (Fig. 5.1). At $\theta = 0$, $\Phi_m = 1.5\, r_{cell} E_e$ and at $\theta = \pi$, $\Phi_m = -1.5\, r_{cell} E_e$. Hence, the difference in the transmembrane potential between the two diametrically opposite points of the cell surface is $3\, r_{cell} E_e$. For a cell of radius $r_{cell} \sim 10$ μm, to produce a transmembrane potential $\Phi_m \sim 0.5$ V, the applied field required is ~300 V/cm.

Electroporation phenomenon depends on various factors such as cell geometry (shape and diameter of cell), nature of buffer used to suspend the cells, temperature, magnitude of applied electric field, and also the cell membrane composition in terms of cholesterol content [31]. Neu et al. gave a very detailed theoretical formulation describing the biophysical basis of cell electroporation [32]. Shil et al. described the physical basis for electroporation in a cell with cholesterol containing membrane [33].

5.4 Combining Radiation with Anticancer Drugs and Electroporation

With the advancement in electronics, many companies are manufacturing electroporators of different specifications. Today, the most frequent application of electroporation is transfection of cells, involving the cellular uptake of DNA molecules. Also, this technology is used for the introduction of fluorescent probes into cells, electroloading of drugs, and transportation of molecules into the cells, etc. The applications have been extended to cancer research as well. This resulted in the emergence of a branch of therapeutics called electrochemotherapy (ECT), where the electric pulses are used to permeabilize the cell membrane and enhance the uptake of the molecules [34–35]. Electrochemotherapy is focused on three aspects: (i) electroporation of cells in living tissues, (ii) potentiation of cytotoxic drugs that are nonpermeant to cells, and (iii) intrinsic response of the body systems, i.e., immune response and blood flow patterns of the patient.

Initial electrochemotherapy had been performed with anticancer drug bleomycin and cisplatin yielded excellent results. As a result, bleomycin has been used for clinical trials [36–40]. Bleomycin was administered by intravenous injection, intratumoral or intraarterial [41, 42]. The application of the drug was mostly of the dose of 10–15 mg/m^2 and the square (or rectangular) wave electric pulses with 100 μs duration and 1300 kV/cm field had been used [35–42]. In most cases, the electric pulses were applied 1–15 min after the intratumoral injection of bleomycin.

Apart from bleomycin, cisplatin was also tried out for electrochemotherapy. Major works were reported by the G. Sersa group [43, 44] involving the use of cisplatin in the treatment of melanoma. Combination studies had been reported involving the electroporation, radiation and cisplatin. In in vivo models, researchers found that electroporation had enhanced the radiation toxicity and the cisplatin induced toxicity to a great extent.

5.5 New Protocols–Radio-Electro-Chemotherapy

Doxorubicin hydrochloride is widely used as an anticancer drug for various types of cancers. Although it is an effective anticancer drug, this anthracyclin derivative is known to be cardiotoxic causing various effects in patients. Hence, our research aimed at enhancing the local potentiation of doxorubicin (DOX) and radiation with electroporation treatment. The aim was to deliver drug and radiation locally and enhance the effects by local electroporation at the tumor.

With this endeavor, we undertook in vitro studies on cell-lines to understand the cellular effects of combined treatment and then undertook in vivo studies on solid tumors in Swiss mice.

5.5.1 Effects on Cancer Cells: In vitro Studies

To study the effects of combination therapy involving radiation, drug and/or electroporation on cancer cells, in vitro studies were performed on murine Ehrlich ascites carcinoma (EAC) cells. The cell line was maintained in vivo in Swiss mice preferably in the peritoneal cavity of mice. Female Swiss mice (6–7 weeks old) weighing ~25 g and reared in the Animal house of the Bhabha Atomic Research Centre (BARC) was used for the study. The guidelines issued by the Institutional Animal Ethics Committee of BARC, Government of India, regarding the care and experimental use of the animals were strictly followed. These rules are in tune with the national laws on animal experimentation, Republic of India.

For the purpose of irradiation, Co-60 gamma rays from a Junior Theratron Teletherapy facility at BARC, Mumbai, was used (dose rate: 0.37 Gy/min). Electroporation was carried out using the indigenously built medical electroporator (BARC-RB&HSD/01-04) with rectangular waves (1–2 kV/cm; duration; 200 μs;

number of pulses/burst: 8; number of bursts: 10). Specially designed cuvettes with parallel-plate electrodes were used for electroporation of the cell suspension. Cellular effects of the various treatments were evaluated in terms of (i) generation of reactive oxygen species (ROS) by fluorescence techniques using 2′-7′-dichlorofluorescein diacetate (H_2DCFDA), (ii) changes in mitochondrial potential using the spectrofluorometric method with 3,3-dihexyloxacarbocyanine iodide ($DiOC_6$) as the fluorescent probe, (iii) membrane oxidative damage by spectrofluometer with 1,6-diphenyl-1,3,5-hexatriene (DPH) as probe, (iv) determination of apoptosis by *Annexin V* [45].

EAC cells were harvested and treated with DOX combining with either electroporation or radiation. It was observed that treatment with 5 μg/mL DOX and 2 Gy Co-60 Gamma rays resulted in 2-fold increase in the cellular ROS (versus control, $p < 0.001$, *t*-test), 2-fold decrease (versus control, $p < 0.001$, *t*-test) in the mitochondrial potential, and a 3-fold increase in apoptotic index (versus control, $p < 0.001$, *t*-test). When EAS cells were treated with 5 μg/mL DOX and electroporation (2 kV/cm; duration; 200 μs; number of pulses/burst: 8; number of bursts: 10), we observed 2.5-fold increase in the cellular ROS (versus control, $p < 0.001$, *t*-test), 3-fold decrease (versus control, $p < 0.001$, *t*-test) in the mitochondrial potential, and a 4-fold increase in apoptotic index (versus control, $p < 0.001$, *t*-test).

Treatment of EAC cells with 10 μg/mL DOX and 2 Gy Co-60 Gamma rays resulted in 2.6-fold increase in the cellular ROS (versus control, $p < 0.001$, *t*-test), 2.3-fold decrease (versus control, $p < 0.001$, *t*-test) in the mitochondrial potential, and a 3.5-fold increase in apoptotic index (versus control, $p < 0.001$, *t*-test). When EAS cells were treated with 10 μg/mL DOX and electroporation (2 kV/cm; duration; 200 μs; number of pulses/burst: 8; number of bursts: 10) we observed 3.2-fold increase in the cellular ROS (versus control, $p < 0.001$, *t*-test), 3.5-fold decrease (versus control, $p < 0.001$, *t*-test) in the mitochondrial potential, and a 5-fold increase in apoptotic index (versus control, $p < 0.001$, *t*-test).

The results implied that combination of DOX with electroporation, that is, electrochemotherapy yielded better cytotoxic effects in EAC cells than the treatment combination of DOX with gamma irradiation.

5.5.2 Effects on Tumors: In vivo Studies

Initial experiments were performed to study the dose response of DOX on murine fibrosarcoma, grown in the right hind leg of Swiss Mice. Animals were treated with doses of DOX at 0.6, 1.2, and 2.4 mg/kg of body weight. Based on the outcome, a DOX dose of 0.6 mg/kg of body weight was chosen for the combined treatment.

Drug was injected intramuscular injections and electroporation carried out within 4 min. For combined modality therapy, drug and electroporation treatments were delivered within 10 min of irradiation with Co-60 gamma rays. Following treatment groups were considered for the study: control (untreated), vehicle control (only 0.9% NaCl), only electroporation, only radiation, DOX + radiation, DOX + electroporation, radiation + electroporation, and radiation + DOX + electroporation.

For all these groups, radiation treatment of Co-60 gamma rays (dose rate 0.37 Gy/min) and electroporation (1 kV/cm, 200 μs, 8 pulses per burst, 10 bursts) from medical electroporator (BARC-RB&HSD/01-04, BARC, Mumbai) was used.

The effects of the treatments were evaluated in terms of tumor growth kinetics. It was observed that individual treatments with Radiation, DOX and electroporation produced tumor growth delay of 0.72, 0.54 and 0.66 days respectively, compared to control. The highest growth delay (1.72 days) was observed in mice treated with radiation + DOX + electroporation (Table 5.1).

Table 5.1 Tumor growth delay in murine fibrosarcoma

Treatment groups	Tumor growth delay
Control	—
Vehicle control	—
Electroporation	0.72
Radiation	0.54
DOX	0.66
DOX + electroporation	1.2
DOX + radiation	1.5
Radiation + DOX + electroporation	1.72

The study demonstrated that the combination of ionizing radiation (Co-60 gamma rays), local delivery of DOX with electroporation has enhanced tumor toxicity. Thus, with a lower dose of DOX and moderate doses of radiation (2 Gy), it was possible to effectively reduce tumor growth [46].

5.6 Conclusion

The use of electroporation has been found to enhance the effects of moderate dose of ionizing radiation and lower dose (half the standard human dose) of anticancer drug DOX. The combination of all three in solid tumors in mice reduced the tumor growth by almost half to that of untreated. Hence, the combination of irradiation followed by local potentiation of anticancer drugs using electroporation, a protocol which can be called *radio-electro-chemotherapy*, holds promise of effective therapy of solid tumors. The studies hold promises for the combination therapy of solid tumors in future.

References

1. Levi, R., Miller, R. A. (1991) Antibiotics in cancer therapy. In: (De Vita VT Jr, Hellman, Rosenberg, S. A., eds), *Biological Therapy of Cancer*. Philadelphia: Lippincott.
2. Steerenberg, P. A., Stom, G., De Groot, G. (1988) Liposomes as drug career system for cis-diamminedichloroplatinum (II). II antitumor activity in vivo, induction of drug resistance, nephrotoxicity and Pt distribution, *Cancer Chemother. Pharmacol.*, **21**, pp. 299–307.
3. Lubbe, A. S., Bergemann, C., Reiss, H., Huhnt, W., Fricke, T., Brick, J., Huhn, D. (1996) Clinical experience with magnetic drug targeting: Phase I trials with 4′ epidoxorubicin in 4 patients with advanced solid tumors, *Can. Res.*, **56**, pp. 4686–4693.
4. Jekunen, A. P., Shalinsky, D. R., Hom, D. K. (1993) Modulation of cisplatin cytotoxicity by permeabilization of plasma membrane by digitonin in vitro, *Biochem. Pharmacol.*, **45**, pp. 2079–2085.
5. Marverti, G., Andrew, P. A., Piccini, G., Ghiaroni, S., Berbeiri, D., Moruzzi, M. (1997) Modulation of Cis-diamminedichloroplatinum (II) accumulation and toxicity by spermine in sensitive and resistant human ovarian carcinoma cells, *Eur. J. Cancer*, **33**, pp. 669–675.

6. Neumann, E., Sowers, A., Jordan, C. (1989) *Electroporation and Electrofusion in Cell Biology*. New York: Plenum press.
7. Kinosita, K., Tsong, Ty. (1977) Formation and resealing of pores of controlled sizes in human erythrocyte membrane, *Nature*, **268**, pp. 438–441.
8. Abidor, I. G., Arakelyan, V. B., Chernomomordik, L. V., Chizmadzhev, Y. A., Pastushenko, V. P., Tarasevich, M. R. (1977) Electrical breakdown of bilayer lipid membranes: I main experimental facts and there qualitative discussion, *Bioelectrochem. Bioenerg*., **6**, pp. 37–52.
9. Mir, L. M., Orlowski, S. (1999) Mechanisms of electrochemotherapy, *Adv. Drug Deliv. Rev.*, **35**, pp. 107–118.
10. Heller, R., Gilbert, R., Jaroszeski, M. (1993) Clinical applications of electrochemotherapy, *Adv. Drug Deliv. Rev.*, **35**, pp. 119–129.
11. Sersa, G., Cemazar, M., Miklavcic, D. (1995) Antitumor effectiveness of electrochemotherapy with cis-diamminedichloroplatinum(II), *Cancer Res.*, **55**, pp. 3450–3455.
12. Mir, L. M., Orlowski, S., Belehradek, J., Paoletti, C. (1991) electrochemotherapy potentiation of antitumor effects of bleomycin by local electric pulses, *Eur. J. Cancer*, **27**, pp. 68–72.
13. Cemazar, M., Miklavcic, D., Scancar, J., Dolzan, V., Golouh, R., Sersa, G. (1999) Increased platinum accumulation in SA-1 tumor cells after in vivo chemotherapy with cisplatin, *Br. J. Cancer.*, **79**, pp. 1386–1391.
14. Sale, A., Hamilton, W. (1967) Effects of high electric fields on microorganisms I, killing of bacteria and yeasts, *Biochem. Biophys. Acta*, **148**, pp. 781–788.
15. Neumann, E., Rosenheck, R. (1972) Permeability changes induced by electric impulses in vascular membranes, *J. Membrane Biol.*, **10**, pp. 279–290.
16. Kinosita, K., Tsong, T. (1977) Formation and resealing of pores of controlled sizes in human erythrocyte membranes, *Nature*, **268**, pp. 438–441.
17. Abidor, I., Arakelan, V., Chernomordik, L., Chimadzhev, Y., Pastuschenko, V., Tarasevich, M. (1979) Electrical breakdown of bilayer membranes I. The main experimental facts and qualitative discussions, *Bioelectrochem. Bioenerg.*, **6**, pp. 37–52.
18. Benz, R., Zimmermann, U. (1980) Relaxation studies on the cell membrane and lipid bilayer in the high electric field range, *Bioelectrochem. Bioeng.*, **7**, pp. 723–739.

19. Benz, R., Conti, F. (1981) Reversible electrical breakdown of squid giant axon membrane, *Biochem. Biophys. Acta*, **645**, pp. 115–123.
20. Chernomordik, L., Sukherev, S., Popov, S., Pastuschenko, V., Sokirko, A., Abidor, I., Chizmadzhev, Y. (1987) The electric breakdown of cells and lipid membranes: The similarity of phenomenologies, *Biochem. Biophys. Acta*, **902**, pp. 360–373.
21. Kinosita, K. Jr, Ashikawa, I., Saita, N., Yoshimura, H., Itoh, H., Nagayama, H., Ikegami, A. (1988) Electroporation of cell membrane visualized under a pulsed laser fluorescence microscope, *Biophys. J.*, **53**, pp. 1015–1019.
22. Hibino, M., Shigemuri, M., Itoh, H., Nageyama, K., Kinosita, K. (1991) Membrane conductance of an electroporated cell analyzed by submicrosec imaging of transmembrane potential, *Biophys. J.*, **59**, pp. 209–220.
23. O'Neill, J. J., Tung, L. (1991) Cell attached patch clamp study of the electropermeabilization of amphibian cardiac cells, *Biophys. J.*, **59**, pp. 1028–1039.
24. Mir, L. M., Banaoun, H., Pauletti, C. (1988) Introduction of definite amounts of nonpermeant molecules into living cells after electropermeabilization. Direct access to the cytosol, *Exp. Cell Res.*, **17**, pp. 15–25.
25. Chakraborti, R., Wylie, D., Schuster, S. M. (1989) Transfer of monoclonal antibodies into mammalian cells by Electroporation, *Biol. Chem.*, **264**, pp. 15494–15500.
26. Lambert, H., Pankov, R., Gauthier, J., Hancock, R. (1990) Electroporation mediated uptake of proteins into mammalian cells, *Biochem. Cell Biol.*, **68**, pp. 729–734.
27. Bartoletti, D. C., Harrison, G. I., Weaver, J. C. (1989) The number of molecules taken up by electroporated cells: Quantitative determination, *FEBS Lett.*, 256, pp. 4–10.
28. Proausnitz, M. R., Lau, B. S., Milano, C. D., Coner, S., Langer, R., Weaver, J. C. (1993) A quantitative study of Electroporation showing a plateau in net molecular transport, *Biophys. J.*, **65**, pp. 414–422.
29. Proausnitz, M. R., Milano, C. D., Gimm, J. A., Langer, R., Weaver, J. C. (1994) Quantitative study of molecular transport due to Electroporation: Uptake of bovine serum albumin by human red blood cell ghosts, *Biophys. J.*, **66**, pp. 1522–1530.
30. Weaver, J. C. (1995) Observation of extremely heterogeneous electroporative uptake which changes with electric field pulse amplitude in *Saccharomyces cerevisiae*, *Biochim. Biophys. Acta*, **1234**, pp. 52–62.

31. Kanduse, R. M., Sentjure, M., Miklavcic, D. (2006) Cell membrane fluidity related to electroporation and resealing, *Eur. Biophys. J.*, **35**(3), pp. 196–204.
32. Smith, K. C., Neu, J. C., Krassowska, W. (2004) Model of creation and evolution of stable electropores for DNA delivery, *Biophys. J.*, **86**(5), pp. 2813–2826.
33. Shil, P., Bidaye, S., Vidyasagar, P. B. (2008) Analysing the effects of surface distribution of pores in cell electroporation for a cell membrane containing cholesterol, *J. Phys. D: Appl. Phys.*, **41**, pp. 551–557.
34. Mir, L., Orlowski, S., Belehradek, J. Jr., Tiesse, J., Rols, M., Sersa, G., Miclavcic, D., Gilbert, R., Heller, R. (1995) Biomedical applications of electric pulses with special emphasis on antitumor electrochemotherapy, *Bioelectrochem. Bioenerg.*, **38**, pp. 203–207.
35. Gehl, J., Skovsgaard, T., Mir, L. (1998) Enhancement of cytotoxicity by Electropermeabilization: An improved method for screening drugs, *Anticancer Drug*, **9**, pp. 319–325.
36. Rodrigues, C., Barroso, B., Almanza, J., et al. (2001) Electrochemotherapy in primary and metastatic skin tumors phase II trials using intralesional bleomycin, *Arch. Med. Res.*, **32**, pp. 273–276.
37. Gehl, J., Geertsen, P. F. (2000) Efficient palliation of haemorrhaging malignant melanoma skin metastasis by electrochemotherapy, *Melanoma Res.*, **10**, pp. 585–589.
38. Rols, M., Bachaund, J., Giraud, P., et al. (2000) Electrochemotherapy of cutaneous metastasis in malignant melanoma, *Melanoma Res.*, **10**, pp. 468–474.
39. Heller, R., Jaroseski, M., Glass, I. (1996) Phase I/II trials for the treatment of cutaneous and subcutaneous tumors using electrochemotherapy, *Cancer*, **77**, pp. 964–971.
40. Sersa, G., Cuferm, T., et al. (2000) Electrochemotherapy with bleomycin in the treatment of hypernephroma metastasis, Ace report and literature review, *Tumori*, **86**, pp. 163–165.
41. Kubota, Y., Mir, L., et al. (1998) Successful treatment of metastatic skin lesions with electrochemotherapy, *J. Urol.*, **160**, p. 1426.
42. Domenge, C., Orlowski, S., Lubonski, B. (1996) Antitumor electrochemotherapy: New advances in the clinical protocol, *Cancer*, **77**, pp. 956–963.
43. Sersa, G., Stabuc, B., Cemazar, M., et al. (2000) Electrochemotherapy with cisplatin: Systemic antitumor effectiveness of cisplatin can

be potentiated locally by the application of electric pulses in the treatment of malignant melanoma metastasis, *Melanoma Res.*, **10**, pp. 381–385.

44. Sersa, G., Stabuc, B., Cemazar, M., et al. (1998) Electrochemotherapy with cisplatin: Potentiation of local cisplatin antitumor effectiveness by application of electric pulses in cancer patients, *Eur. J. Cancer,* **3**, pp. 1213–1218.
45. Shil, P., Sanghavi, S. H., Vidyasagar, P. B., Mishra, K. P. (2005) Enhancement of radiation cytotoxicity in murine cancer cells by electroporation: In vitro and in vivo studies, *J. Environ. Pathol. Toxicol. Oncol.,* **24**, pp. 291–298.
46. Shil, P., Kumar, A., Vidyasagar, P. B., Mishra, K. P. (2006) Electroporation enhances radiation and doxorubicin-induced toxicity in solid tumor in vivo, *J. Environ. Pathol. Toxicol. Oncol.*, **25**, pp. 625–632.

Chapter 6

Motivation to Explore New Techniques for Synthesis of Metal Nanoparticles and Their Immense Importance in Biological and Medicinal Applications

Kashinath A. Bogle,[a] Vasant N. Bhoraskar,[b] Sanjay D. Dhole,[b] Megha P. Mahabole,[a] and Rajendra S. Khairnar[a]

[a]*Thin Films and Sensors Laboratory, School of Physical Sciences, Swami Ramanand Teerth Marathwada University, Nanded, Maharashtra, India*
[b]*Microtron Laboratory, Department of Physics, Savitribai Phule Pune University, Pune, Maharashtra, India*

kashinath.bogle@gmail.com

Nanoparticles of metals with size ranging from 1 to 100 nm have unique properties that assist in molecular diagnostics and therapies. Various methods including physical and chemical routes are available for synthesis of metal nanoparticles. In addition to these methods, a new radiation-assisted synthesis method discovered provides a feasible alternative. In this work, silver and gold nanoparticles have been synthesized by irradiating solutions, prepared by mixing appropriate metal salt in 2-propanol and distilled water having PVA, with 6 MeV electrons. The effect of

Radiation in Medicine and Biology
Edited by Pandit B. Vidyasagar, Sagar S. Jagtap, and Omprakash Yemul
Copyright © 2017 Pan Stanford Publishing Pte. Ltd.
ISBN 978-981-4745-92-5 (Hardcover), 978-1-315-20656-1 (eBook)
www.panstanford.com

electron fluence on the particle size is described, at room temperature. Metal nanoparticle solutions were characterized by UV-vis absorption spectroscopy, X-ray diffraction, transmission electron microscopy, and energy dispersive spectroscopy techniques. The study reveals that the particle size decreases and monodispersivity increases with increasing electron fluence, which is a new finding in this work. This method is easy in operation and even a period of about 10 min is sufficient to obtain a signature of the nanoparticles. In most of the medical applications, silver and gold nanoparticles antimicrobial property was majorly explored, though the anti-inflammatory property has its fair share of applications.

6.1 Introduction

Despite the notorious nature of radiation, at a different level, it plays an important role in the day-to-day life, such as microwave radiation for instant heating of food and radiation treatment for the preservation of food from microorganisms, bacteria, viruses, etc.

High-energy radiation (electron, gamma, position, X-rays) has revolutionized the medical field by enhancing the ability of medical professionals to treat and diagnose different type of diseases in humans as well as animals. Today, CT scans, magnetic resonance imaging (MRI)/positron emission (PE) tomography and use of X-rays along with advanced computer technology produce highly detailed horizontal/axial/cross-sectional images of the body, including the bones, muscles, fat, organs, and blood vessels. Using these scans, medical professionals can accurately find clots and other harmful medical conditions of the patients.

Living organisms in human or animals are made up of cells of size typically ~10 μm, which have much smaller parts of sub-micron sizes. Proteins are even smaller with a typical size of just 5 to 10 nm and are comparable to the dimensions of smallest manmade marvel "nanoparticles." Prof. O. V. Salata from the University of Oxford, UK, states that this simple size comparison gives an idea of using nanoparticles as very small probes that would allow us to spy at the cellular machinery without introducing too much interference. Understanding of biological processes on the nanoscale level is a strong driving force behind the development of nanotechnology [1].

The medical properties of metals are known from ancient Ayurvedic treatments. This promotes the use of herbs and herbal preparation routes for the treatment of various illnesses. The system frequently employs unique metallic preparation known as *bhasmas* with extracts of herbal juices for curing diseases. These *bhasmas* contain herbo-metal particles of sizes 10–50 nm, and use of specific nanoparticles of metal for the treatment of diseases has been a common practice in Ayurveda [2]. Some of the common metals used are silver (Ag) and gold (Au). In modern nanomedicine and Ayurvedic treatment, Ag nanoparticles are used to cure ailments such as muscle wasting, nerve disorders, brain diseases, and meningitis and Au nanoparticles have shown special cure for rheumatoid arthritis [2].

In addition to the above-mentioned medical applications, metal nanoparticles are used as antimicrobial agents in many public places such as railway stations and elevators around the world and they show good antimicrobial action due to their high toxicity to microorganisms, including 16 major species of bacteria [3, 4].

There are various physical and chemical routes available for synthesizing metal nanoparticles of desired size and shapes. This work summarizes a new method developed for the synthesis of metal nanoparticles, where high-energy notorious radiation (6 MeV electron) is used to obtain monodispersed metal nanoparticles [5]. These prepared nanoparticles can be utilized for said the already-mentioned biological and medicinal applications.

6.2 Experimental

Nanoparticles of metals such as silver and gold were synthesized in the aqueous medium using the 6 MeV electron irradiation technique. In this technique, silver nitrate ($AgNO_3$) and chloro-auric acid ($HAuCl_4$) were used for synthesizing Ag and Au nanoparticles. The particles formed in the solution were stabilized by adding PVA in aqueous solution.

During the synthesis of Ag nanoparticles, samples of Ag-solution were prepared by mixing 3 wt% of polyvinyl alcohol

(PVA), 0.001 M $AgNO_3$ and 2-propanol in distilled water. With the presence of alcohol, such as 2-propanol in the solution, the $OH^•$ and $H^•$ radicals are scavenged to yield reducing organic radicals [5]. The solution was prepared in dark room and precautions were taken not to allow Ag to reduce at room temperature. The Ag-solution was filled in different thin-walled glass bottles each of size 15 mm diameter and 40 mm height. These glass bottles were covered with thin aluminum foil to avoid a direct exposure to day light.

The Microtron accelerator was tuned to obtain an electron beam of energy ~6 MeV energy, flux $\sim 10^{12}$ e/cm^2-s, with a pulse width of 1.6 μs at a repetition rate of 50 PPS. The Ag-solution was irradiated uniformly with 6 MeV electrons to desired electron fluence in the range from 2×10^{13} e/cm^2 to 4×10^{13} e/cm^2.

By following the same experimental procedure Au, particles were also synthesized. The samples of Au-solution were made in distilled water by mixing 0.001 M $HAuCl_4$ and 3 wt% of PVA and 2-propanol. The solution in the glass bottle was irradiated and electron fluence was varied in the range from 1×10^{14} to 1×10^{15} e/cm^2.

The synthesized metal nanoparticles were characterized using different techniques such as X-ray diffraction (XRD), UV-Vis absorption spectroscopy, transmission electron microscopy (TEM) and energy dispersive spectroscopy (EDS). For the XRD analysis, a few drops of solution were spread on quartz plate and dried. For the TEM measurements, a self-supported ultrathin film of the respective electron-irradiated solution was made in the laboratory. Then with the help of a fine-edge blade, a small edge piece was cut from this ultrathin film and mounted on the TEM grid. The optical absorption spectra of the electron-irradiated Ag and Au solutions were measured using a UV-Vis spectrophotometer.

6.3 Results and Discussion

The recorded XRD spectra of the thin coatings casted from the Ag-solution irradiated at electron fluences 2×10^{13} and 3×10^{15} e/cm^2 are shown in Figs. 6.1a,b, respectively. Four distinct XRD peaks at 2θ values of 38.40, 44.30, 64.40, and 77.40

represent the (111), (200), (220), and (311) crystalline planes of the face-centered cubic Ag crystal structure, respectively [6–8]. The sizes of the particles were estimated by analyzing the XRD spectra using Scherrer's formula [9]. The range of size distribution estimated from the XRD spectra shown in Figs. 6.1a,b are 57 to 64 nm and 8 to 12 nm, respectively. At a lower electron fluence (2×10^{13} e/cm^2), the average size of the Ag nanoparticles is 60 nm and it decreased to 10 nm at an electron fluence of 3×10^{15} e/cm^2. These results indicate that the size of Ag nanoparticle decreases with increasing electron fluence from 2×10^{13} to 3×10^{15} e/cm^2.

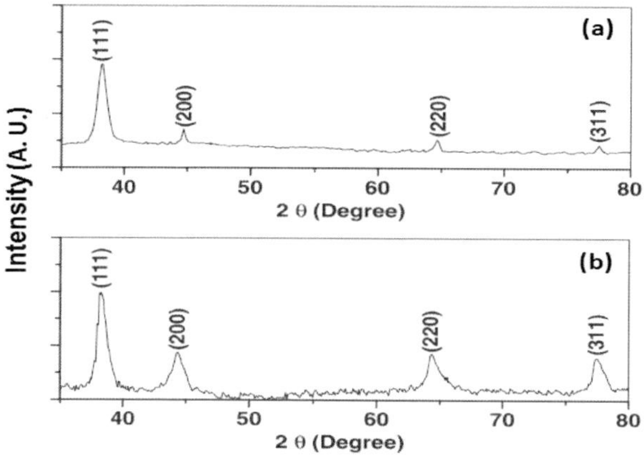

Figure 6.1 The XRD spectra of the Ag-solution irradiated at an electron fluence of (a) 2×10^{13} e/cm^2 and (b) 3×10^{15} e/cm^2.

In case of Au nanoparticles, the XRD spectra recorded for the thin coatings made from the electron-irradiated Au-solution at an electron fluences of 1×10^{14} and 1×10^{15} e/cm^2 are shown in Fig. 6.2. The diffraction peaks that appeared at $2\theta \sim 38.2°$, 44.3°, 64.6°, and 77.4° are assigned to the (111), (200), (220), and (311) planes of the face-centered cubic phase of Au crystal, respectively [10, 11]. A broad nature of XRD peaks in Fig. 6.2a indicates the formation of nano-size Au crystallites. These results indicate that smaller-size Au nanoparticles are produced at higher electron fluences (10^{15} e/cm^2). The particle size distribution was found to vary from 18 to 12 nm and 4 to 2 nm for the particles synthesized in Au-solution irradiated at an electron fluence of 1×10^{14} and 1×10^{15} e/cm^2. These results therefore clearly show that Au nanoparticles can be synthesized by 6 MeV electron irradiation methods and the size of the Au nanoparticles decreases with increasing electron fluence.

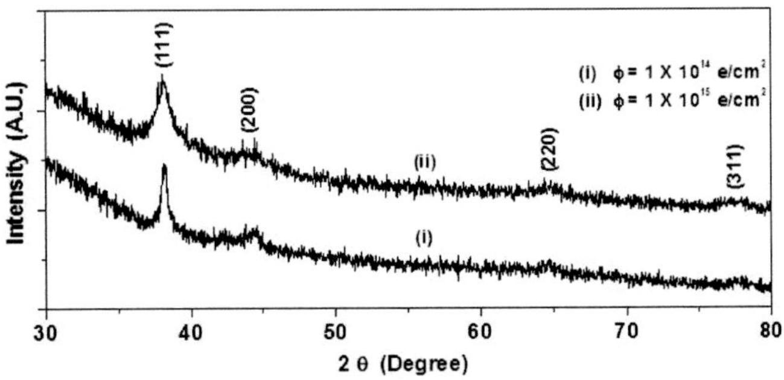

Figure 6.2 The XRD spectra of Au-solution irradiated at an electron fluence of (i) 1×10^{14} e/cm^2 and (ii) 1×10^{15} e/cm^2.

The decrease in the size of nanoparticle with increasing electron fluence as evidenced from the XRD spectra was further confirmed by the UV-Vis absorption spectroscopy measurements. Based on the Mie theory [12], it is considered that the absorption peak originating from the surface plasmon resonance get shifted to lower wavelength side with decreasing particle size, and towards higher wavelength with increasing particle size. It is well known that the position and shape of the plasmon absorption of metal nanoclusters are strongly dependent on the particle size, dielectric medium, and surface-adsorbed species [13–15]. By keeping all these parameters under consideration, the medium (solvent) and surface adsorbed species (PVA) in the solution were kept the same. Using a HITACHI 330 made spectrophotometer, the optical absorption properties of the electron-irradiated solutions were studied. The optical absorption spectra for the Ag-solution irradiated with electrons at fluences in the range 2×10^{13} to 3×10^{15} e/cm^2 are shown in Fig. 6.3. The absorption spectrum for each electron-irradiated samples was measured with reference to the solvent (water) used for the synthesis.

The optical absorption spectrum for the Ag-solution irradiated at an electron fluence of 2×10^{13} e/cm^2 is dominated by a single absorption peak at ~455 nm, which corresponds to the surface plasmon resonance peak of Ag nanoparticles [15].

At higher electron fluence, 3 × 10^{16} e/cm^2, this peak shifts to lower wavelength ~394 nm. This blue shift of ~61 nm in the plasmon absorption peak reveals that the size of the particle reduces with increasing electron fluence.

Similarly, the optical absorption spectra of the Au-solution irradiated at various electron fluences in the range from 1 × 10^{14} to 1 × 10^{15} e/cm^2 are shown in Fig. 6.4.

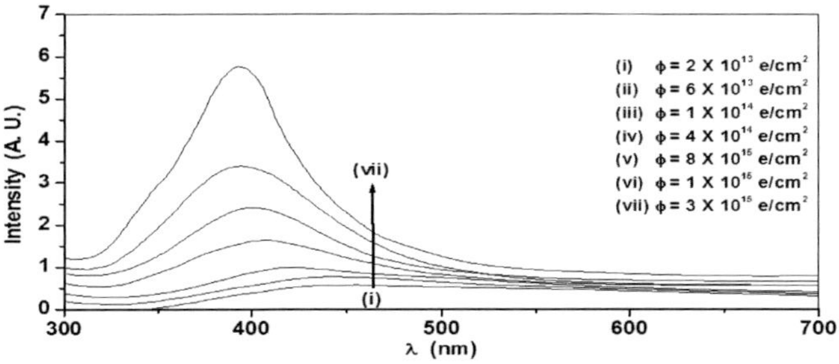

Figure 6.3 The optical absorption spectra of Ag-solution irradiated at different electron fluence over the range 2 × 10^{13} to 3 × 10^{15} e/cm^2.

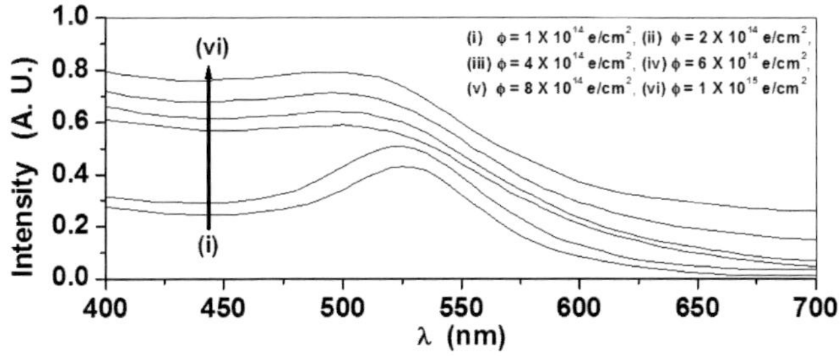

Figure 6.4 The optical absorption spectra of Au-solution irradiated at different electron fluences over the range from 2 × 10^{13} to 3 × 10^{15} e/cm^2.

The results are the same as obtained in the case of Ag nanoparticles (Fig. 6.3). The surface plasmon absorption peak for the Au nanoparticles synthesized at an electron fluence of 1 × 10^{14} e/cm^2 is observed at ~531 nm, which is the typical surface plasmon absorption peak for spherical Au nanoparticles [10–18]. The plasmon absorption peak for Au nanoparticles blue shifted (~38 nm) to a lower wavelength from 531 to 493 nm, with increasing electron fluence up to 1 × 10^{15} e/cm^2.

A typical TEM image of a thin film made from the Ag-solution irradiated with 6 MeV electrons at different fluencies is shown in Fig. 6.5. At lower electron fluence of 2×10^{13} e/cm^2, the particles are of spherical shape, but with a high degree of polydispersivity and a wide size distribution.

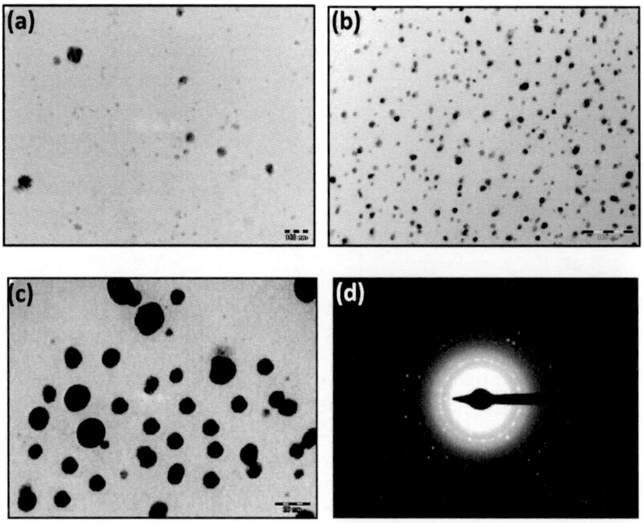

Figure 6.5 TEM image of the Ag nanoparticles synthesized at an electron fluence of (a) 2×10^{13} e/cm^2, (b) 3×10^{15} e/cm^2, (c) magnified image of Ag nanoparticles in (b) and (d) the electron diffraction pattern of the (b) sample.

The size of the Ag nanoparticle varied from 40 to 80 nm, with average size of ~60 nm. Moreover, it can be seen in Fig. 6.5a that the Ag nanoparticles of lower size, ~20 nm, are also present in the sample, but in lower number. These results indicate that the Ag particles of relatively large size are produced at lower electron fluence.

Figures 6.5b,c reveal that Ag nanoparticles are nearly spherical in shape, which is consistent with the result obtained from optical absorption spectroscopy, as shown in Fig. 6.3. In addition to this, the particle size distribution in this sample lies in the range of 10 to 15 nm, with average size ~12 nm. The crystal structure of the Ag nanoparticles was again confirmed using electron diffraction. The obtained electron diffraction pattern for the Ag nanoparticles synthesized at an electron fluence of

3×10^{15} e/cm^2 is shown in Fig. 6.5d. The diffraction rings shown in Fig. 6.5d could be indexed to (111), (200), (220), and (311) planes of face-centered cubic phase of Ag crystals [6, 19]. These results are consistent with those of the XRD peaks shown in Fig. 6.1.

Similarly, the Au nanoparticles synthesized in the Au-solution were also studied by the TEM technique, and the results are shown in Fig. 6.6.

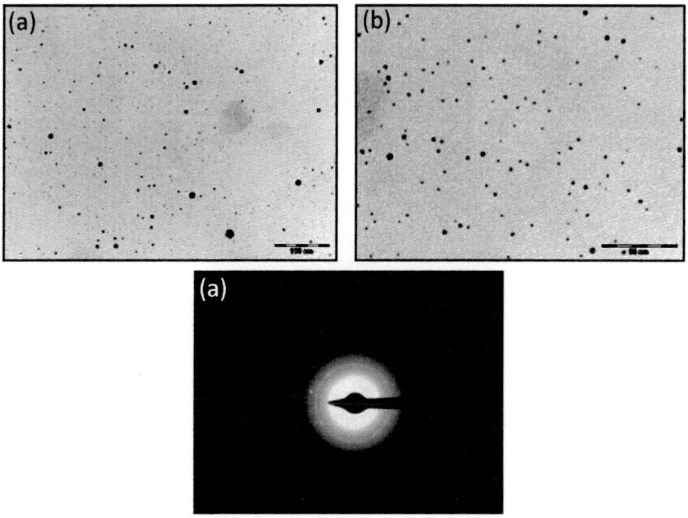

Figure 6.6 TEM images of Au nanoparticles synthesized at electron fluence: (a) 1×10^{14} e/cm^2 and (b) 1×10^{15} e/cm^2, and (c) electron diffraction pattern for (b) sample.

For Au nanoparticles synthesized at an electron fluence 1×10^{14} e/cm^2, a wide distribution of particle size was observed, which may vary from 15 to 5 nm with an average particle size ~10 nm. Moreover, there are very less number of particles of size ~5 nm but a large number of particles with average size around ~10 nm. For Au nanoparticles synthesized at 1×10^{15} e/cm^2, as shown in Fig. 6.6b, the particle size distribution decreased with an average particle size ~3 nm. As seen in the TEM image, most of the isolated particles have nearly spherical shape. The electron diffraction pattern taken for the Au nanoparticles synthesized at an electron fluence of 1×10^{15} e/cm^2 shown in Fig. 6.6c could be indexed to the face-centered cubic phase with (111), (200),

(220), and (311) planes of Au crystals [19]. These results are consistent with the XRD spectra shown in Fig. 6.2a.

The presence of metal (Ag and Au) nanoparticles in the thin coatings made from the electron-irradiated solutions was confirmed using an elemental analysis technique such as EDS. The recorded EDS spectra for the PVA-stabilized Ag and Au nanoparticles synthesized at respective electron fluence of 3×10^{15} and 1×10^{15} e/cm^2 are shown in Fig. 6.7.

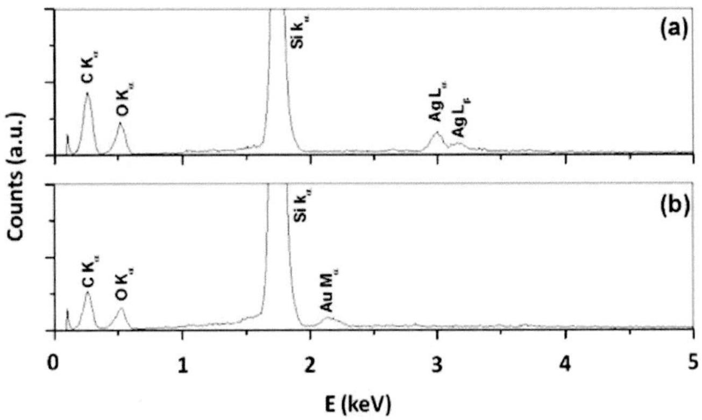

Figure 6.7 EDS spectra for the thin coatings made from the (a) Ag-solution-I, (b) Au-solution-I, and (c) Cu-solution-I irradiated at 1×10^{15} e/cm^2 of 6 MeV electrons.

As explained in experimental section, few drops of respective nanoparticles solution was spread on quartz plate and dried in oven at 60°C. This sample was used for EDS analysis. Ag Lα line at E = 2.983 keV reveals presence of Ag in the sample. Moreover, in case of Au and Cu nanoparticles by the presence of Au Mα line at E = 2.121 keV and Cu Mα line at E = 0.82 keV along with the other line as explained above. The observed core level x-ray line indicates that the respective sample contain metallic Ag, Au and Cu.

The results of XRD, UV-Vis absorption spectroscopy, TEM, and SEM clearly indicate the formation of Ag, Au, and Cu nanoparticles in the respective solutions under 6 MeV electron irradiation. The XRD and optical absorption spectroscopy measurements evidence the formation of relatively large-size metal nanoparticles at lower electron fluence of ~10^{14} e/cm^2. Moreover, small-size metal nanoparticles are formed with increasing electron fluence to higher values of 10^{15} e/cm^2. As shown in Figs. 6.3 and 6.4, the surface plasmon peak for corresponding nanoparticles shifts towards lower wavelength side with increasing electron fluence. This indicates that the size of the nanoparticle decreases with increasing electron fluence and depends on electron fluence. As a matter of fact, this effect can be attributed to the zero valance M^0 atoms produced initially in the solution due to radiolytic reduction of metal salt, water, and PVA under 6 MeV electron irradiation. In addition to this, the irradiation cross-linking of PVA chains, which is used as a particle stabilizer, is also greatly affected by electron fluence. At lower electron fluence, the concentration of the zero valence M^0 atoms is high but there is a low yield of cross-linked PVA chains. As the reaction proceeds, these metal atoms (M^0) act as a nucleation center for the ions present in the solution and further coalescence. At lower electron fluence, a few nucleation centers are produced that can absorb the excess ions, which are produced in the solution. Therefore, lower electron fluence may favor the formation of large-size metal nanoparticles because of low nucleation centers and lower cross-linked density of PVA, resulting in a slow stabilizing rate. This results in growth in the size of the nanoparticles; so there is low concentration of nanoparticles but with large size. With increasing electron fluence, the large-size Ag nanoparticles might be broken into smaller pieces to form small-size nanoparticles or there would be an increase in the number of nucleation centers, but at the same time the rate of production of excess ions remains the same. Moreover, at higher electron fluences the growth rate of the metal nanoparticles is confined by a polymer matrix with a high density of cross-linked polymer [20], resulting in a large number of small-size metal nanoparticles. From XRD and TEM results, shown in Figs. 6.1, 6.2, and 6.5, 6.6, respectively, it is observed that the increase in the FWHM of the XRD peak is related with an increase in the particle size and this decrease in the particle size is observed in the TEM images.

The results shown in this work clearly indicate that electron fluence plays an important role in controlling the size of metal nanoparticles. The literature survey shows that the surface plasmon peak strongly depends on the particle size under identical conditions of the precursor. The increase in electron fluence shows a major blue shift in the surface plasmon absorption peak position of the metal nanoparticles compared with that in the metal nanoparticles synthesized with varying the concentration of PVA in the solution at the same electron fluence.

To understand nucleation and growth of metal nanoparticles in solution, it is necessary to know interaction of high-energy electrons with the medium. Energetic electrons deposit energy in a medium through various processes, among which the prominent processes are ionization and excitation of atoms, breaking of chemical bonds, cross-linking, and disintegration of molecules [21]. During electron irradiation, the following electron-induced radiolytic reactions and disintegration processes can be initiated in an aqueous mixture of the metal salt, PVA, and 2-propanol solution [6, 7, 20–27].

$$n \cdot H_2O \sim\sim\rightarrow \text{e-aq}, H^{\bullet}, OH^{\bullet}, H_2, H_3O^{+}, H_2O_2, \ldots$$

$$MS \rightarrow M^{x+},$$

where MS is the metal salt and M^{x+} is the metal ion produced in the solution due to dissociation of the metal salt under electron irradiation. The metal salts used here are $AgNO_3$ and $HAuCl_4$, which release metal ions such as Ag^+ and Au^{3+} in the solution. The oxidation of hydroxyl radical $OH^{\bullet}$ is avoided by addition of radical scavengers, such as 2-propanol, which yields after reaction with $OH^{\bullet}$ and $H^{\bullet}$ to a secondary reducing radical $(CH_3)_2C^{\bullet}OH$ [6, 20, 27].

$$(CH_3)_2CHOH + OH^{\bullet} \rightarrow (CH_3)_2C^{\bullet}OH + H_2O$$

$$(CH_3)_2CHOH + H^{\bullet} \rightarrow (CH_3)_2C^{\bullet}OH + H_2$$

For the M^+ ions, the electron capture cross section is very high [20] and, therefore, a large number of neutral (M^0) metal atoms can be produced in the solution through the following reaction:

$$M^+ + e^-aq \rightarrow M^0$$

The neutral M^0 atoms can encounter the excess M^+ ions; then they tend to associate with ions and coalesce into dimers. These dimers progressively grow into oligomers followed by a large cluster, which eventually precipitates out in the solution through the following reactions:

$$M^0 + M^+ \rightarrow M^{+2},$$

$$Mn^{-1} + M^+ \rightarrow M^{n+}$$

Moreover, to obtain a stable cluster of nanometeric dimension, the coalescence has to be confined by adding a polymeric molecule acting as a particle stabilizer. Usually polyvinyl alcohol (PVA) is used as a stabilizing agent because it does not reduce ions before irradiation [6, 20]. Moreover, on irradiation with electrons, the process of cross-linking of PVA and subsequently hydrogel formation occurs in the aqueous solution. The higher stability and uniform distribution of metal nanoparticles in a gel matrix, without aggregation of small particles are some of the major advantages offered by such stabilizing polymers [20].

After addition of the particle stabilizer PVA, under electron irradiation the reaction leads to further reduction of M^+ ions through the strongly reducing hydrated electrons as well as by the cross-linked polymeric radicals $PVA^\bullet$ [20]. During electron irradiation, the following disintegration can occur with PVA:

$$PVA(H) + OH^\bullet \rightarrow PVA^\bullet + H_2O$$

$$Mn^+ + e^-aq/PVA^\bullet \rightarrow (M)_n \text{ (Metal nanoparticle)}$$

In this manner, during electron irradiation, the size of the metal particle can grow, and subsequently metal particles of nanometeric dimensions can be synthesized in the solution.

6.4 Conclusion

The results of the present study unambiguously illustrate that the 6 MeV electron irradiation method can be used for synthesizing metal (Ag and Au) nanoparticles of variable sizes at room temperature. The TEM results show that the metal nanoparticles have spherical shape, with face-centered cubic geometry. The size of Ag and Au nanoparticles could be tailored from 60 to 10 nm for Ag and 15 to 3 nm for Au nanoparticles, by varying electron fluence from 10^{13} to 10^{15} e/cm^2. The unique feature of the electron irradiation method is that the particle size can be decreased with increasing electron fluence. In chemical and other methods, the particle size normally increases with increasing the processing period. The metal particles synthesized with the present method were stable even after storing at room temperature for a period of more than six months; therefore they can be utilized biological and medical applications.

References

1. O. V. Salata, Applications of nanoparticles in biology and medicine, *J. Nanobiotechnol.*, 2 (2004) 3.
2. S. Paul and A. Chugh, Assessing the role of Ayurvedic "bhasms" as ethno-nanomedicine in the metal based nanomedicine patent regime, *J. Intellect. Property Act*, 16 (2011) 509.
3. S. Prabhu and E. K. Poulose, Silver nanoparticles: Mechanism of antimicrobial action, synthesis, medical applications, and toxicity effects, *Int. Nano Lett.*, 2 (2012) 32.
4. J. Choi, J. Yoo, M. Lee, E.-G. Kim, J. S. Lee, S. Lee, S. Joo, S. H. Song, E.-C. Kim, J. C. Lee, H. C. Kim, Y.-G. Jung, and S. Kwon, A rapid antimicrobial susceptibility test based on single-cell morphological analysis, *Sci. Transl. Med.*, 6 (2014) 267.
5. K. A. Bogle, S. D. Dhole, and V. N. Bhoraskar, Silver nanoparticles: Synthesis and size control by electron irradiation, *Nanotechnology*, 17 (2006) 3204.
6. M. K. Temgire and S. S. Joshi, Optical and structural studies of silver nanoparticles, *Radiat. Phys. Chem.*, 71 (2004) 1039.
7. M. Kumar, L. Varshney, and S. Francis, Radiolytic formation of Ag clusters in aqueous polyvinyl alcohol solution and hydrogel matrix, *Radiat. Phys. Chem.*, 73 (2005) 21.
8. I. Hussain, B. Mathias, A. J. Papworht, and A. I. Cooper, Stabilized gold and silver hydrosols and gold-polymer composite films, *Langmuir*, 19 (2003) 4831.
9. B. D. Cullity, *Elements of X-Ray Diffraction* (Addison-Wesley, New York, 1959).
10. S. Chen, Y. Liu, and G. Wu, Stabilized and size-tunable gold formed in a quaternary ammonium-based room-temperature ionic liquid under γ-irradiation, *Nanotechnology*, 16 (2005) 2360.
11. W. Chen, W. P. Cai, C. H. Liang, and L. D. Zhang, Synthesis of gold nanoparticles dispersed within pores of mesoporous silica induced

by ultrasonic irradiation and its characterization, *Mater. Res. Bull.*, 36 (2001) 335.

12. G. Mie, Contributions to the optics of turbid media, particularly of colloidal, *Ann. Phys.*, 25 (1908) 377.
13. P. V. Kamat, M. flumiani, and G. V. Hartland, Picosecond dynamics of silver nanoclusters. Photoejection of electrons and fragmentation, *J. Phys. Chem. B*, 102 (1998) 3123.
14. A. Heilmann, A. Kiesow, M. Gruner, and U. Kreibig, Optical and electrical properties of embedded silver nanoparticles at low temperatures, *Thin Solid Films*, 175 (1999) 343.
15. R. He, X. Qian, J. Yin, and Z. Zhu, Preparation of polychrome silver nanoparticles in different solvents, *J. Mater. Chem.*, 12 (2002) 3783.
16. S. S. Joshi, S. F. Patil, V. Iyer, and S. Mahumuni, Radiation induced synthesis and characterization of copper nanoparticles, *Nanstruct. Mater.*, 10 (1998) 1135.
17. T. K. Sau, A. Pal, N. R. Jana, Z. L. Wang, and T. Pal, Size controlled synthesis of gold nanoparticles using photochemically prepared seed particles, *J. Nanopart. Res.*, 3 (2001) 257.
18. H. Abe, K. P. Charle, B. Tesche, and W. Schulze, Surface plasmon absorption of various colloidal metal particles, *Chem. Phys.*, 68 (1982) 137.
19. A. V. Singh, B. M. Bandgar, M. Kasture, B. L. V. Prasad, and M. Sastry, Synthesis of gold, silver and their alloy nanoparticles using bovine serum albumin as foaming and stabilizing agent, *J. Mater. Chem.*, 15 (2005) 5115.
20. W.-T. Wu, Y. Wang, L. Shi, Q. Zhu, W. Pang, G. Xu, and F. Lu, In situ formation of Ag flowerlike and dendritic nanostructures in aqueous solution and hydrolysis of an amphiphilic block copolymer, *Nanotechnology*, 16 (2005) 3017.
21. P. S. Alegaonkar, A. B. Mandle, S. R. Sainkar, and V. N. Bhoraskar, Refractive index and dielectric constant of the boron and fluorine diffused polyimide, *Nucl. Instr. Meth. B*, 194 (2002) 281.
22. Y. Sun and Y. Xia, Shape-controlled synthesis of gold and silver nanoparticles, *Science*, 298 (2002) 2176.
23. W. I. Wu, W. M. Pang, G. Y. Xu, L. Shi, Q. R. Zhu, Y. S. Wang, and F. Lu, In situ formation of Ag flowerlike and dendritic nanostructures in aqueous solution and hydrolysis of an amphiphilic block copolymer, *Nanotechnology*, 16 (2005) 2048.

24. P. Ulanski, E. Bothe, J. M. Rosiak, and C. V. Sonntag, OH-radical-induced crosslinking and strand breakage of poly(vinyl alcohol) in aqueous solution in the absence and presence of oxygen. A pulse radiolysis and product study, *Macromol. Chem. Phys.*, 195 (1994) 1443.

25. A. Henglein, Reduction of $Ag(CN)_2$-on silver and platinum colloidal nanoparticles, *Langmuir*, 17 (2001) 2329.

26. A. Henglein, Radiolytic preparation of ultrafine colloidal gold particles in aqueous solution: Optical spectrum, controlled growth, and some chemical reactions, *Langmuir*, 15 (1999) 6738.

27. A. Henglein, Colloidal palladium nanoparticles: Reduction of Pd(II) by H_2; Pd core Au shell Ag shell particles, *J. Phys. Chem. B*, 104 (2000) 1206.

Chapter 7

Gold Nanoparticle–Assisted Radiation Therapy

Prabhakar Dongre

Department of Biophysics, University of Mumbai, Vidyanagari Santacruz (E), Mumbai, Maharashtra, India

drpmdongre@yahoo.co.in

Nanotechnology is a rapidly growing field with ramifications in medicine, chemistry, biology, engineering, and materials science. Nanomaterials possess unique optical, electrical, and magnetic properties that are different from bulk material of the same chemical. As a result, they are being developed for various applications in medicine such as targeted drug delivery, diagnostics, and therapy. Various metallic nanoparticles are being tested for medical applications. Among them, gold nanoparticles are considered to be biologically safe and biocompatible materials and have attracted considerable attention in cancer imaging and radiation therapy (RT). RT is one of the modalities used in treating cancer. In RT the dose delivered by ionizing radiation can be enhanced in the presence of radio-modifying materials (e.g., high $-z$ number) in cancer cells. Therefore, it is possible to eliminate cancer cells from the tumor. This chapter discusses the effectiveness of gold nanoparticle-assisted radiation therapy.

Radiation in Medicine and Biology
Edited by Pandit B. Vidyasagar, Sagar S. Jagtap, and Omprakash Yemul
Copyright © 2017 Pan Stanford Publishing Pte. Ltd.
ISBN 978-981-4745-92-5 (Hardcover), 978-1-315-20656-1 (eBook)
www.panstanford.com

7.1 Introduction and Background

Radiation therapy (RT) is the clinical and scientific discipline in which high-energy ionizing radiation (photons, electron, protons and heavy charged particles) is used to eliminate masses of cancer cells (or malignant neoplasms), through mitotic death or apoptosis. Almost 52% of cancer patient undergo radiotherapy at least once during their treatment course.

Ionizing radiations' interaction with human tissues occurs at the atomic level through excitation and, more commonly, ionization. When an atom is ionized, its chemical binding properties are altered. If the atom is part of a large molecule, ionization may result in the breakage of the molecule or change in the location of the atom within the molecule. These alterations may impair function and result in cell death. However, cells and tissues can repair, regenerate, and recover. Early effects of radiation are injuries that occur within minutes, hours, and days, while the late effects are those injuries that occur within months, years, and decades after exposure. Radiation acts on a biological system indirectly and directly through the process of ionization and free radical production. An indirect effect occurs as a result of the radiolysis of water and the production of free radicals, while the direct effect occurs when the ionizing radiation interacts directly with a particularly radiosensitive molecule, i.e., deoxyribonucleic acid (DNA). There is strong circumstantial evidence that DNA is the principal target for the biological effects of radiation, including cell killing, carcinogenesis, and mutation. The extent of damage is the direct result of the amount of energy deposited per unit mass, called absorbed dose, which is measured in Gray (Gy).

The objective of radiation therapy is to deliver a prescribed dose to the tumor while limiting the dose to surrounding normal tissues and critical organs. While the prescribed doses of radiation are designed to cause tumor regression, they must also minimize the inevitable consequences of subjecting healthy tissue to radiation. Cancer cells are more susceptible to radiation than most normal human cells, due in part to the higher rate of tumor cell replication and the irregular structure of cancer cell DNA. Still, damaging healthy tissue can inhibit proper physiological functions and lead to a reduction in the patient's quality of life. Thus, the primary goal of RT is to maximize the probability of tumor control while at the same time minimize complication to the normal tissue. In practice, however, we have been hampered by our abilities to identify cancer cells and irradiate them meticulously with radiation without harming the normal tissues. Overcoming this limitation to achieve the goal of radiation therapy is the key component that drives the development in technology, radiation physics, and tumor biology. RT can be delivered by external beam radiation therapy (source of radiation placed far away from the patient) or brachytherapy (source of radiation placed near or within the tumor).

Over the past decade, enormous progress has been made in RT treatment planning and delivery technique, such as three-dimensional conformal radiation therapy (3DCRT), stereotactic radiosurgery (SRS), stereotactic radiotherapy (SRT), intensity-modulated radiotherapy (IMRT), intensity-modulated radiosurgery (IMRS), and, recently, image-guided radiotherapy (IGRT). This was realized primarily due to the groundbreaking innovations in physics and technology, introduction of fast computers, availability of multimodality imaging devices (CT, MRI, PET-CT, etc.), more sophisticated, accurate and robust dose computational algorithms, technological development in computer-controlled treatment delivery machine (linear accelerator), and better understanding of tumor biology.

Heavy elements (high *Z* number) possess higher photoelectric absorption cross sections and they can be potent radiosensitizers [1]. The photoelectric phenomenon takes place below the electron rest energy of 511 keV, beyond which inelastic Compton scattering becomes more prevalent. Gold nanoparticles have larger atomic radius than gold atoms (0.17 nm) and thus should have stronger radiation enhanced effect than the gold atom. Ample literature is available on the mechanisms of interactions of metallic gold nanoaparticles and other nanoparticles with biological systems [2–5]. GNPs are considered biologically safe materials, biocompatible both in vitro and vivo as these do not cause cellular toxicity. They exhibit greater CT attenuation at clinically relevant imaging energies that typical iodine base agent. GNPs can be easily manipulated in a variety of shapes and sizes and possess easily controlled surface chemistry, which allows functions with biological molecules to improve stability and tumor targeting. GNPs absorb high ionizing radiation than soft tissues, which significantly enhances the radiation effect. The biodistribution of GNPs can be imaged before a therapeutic dose is delivered and used for treatment planning and quantified prediction of dose enhancement.

7.2 Gold Nanoparticle–Assisted Radiation Therapy (GNRT)

As mentioned previously, the biological effects of radiation result from both direct and indirect action of radiation. The direct action is based on the interaction between radiation particles and complex body cell molecules (e.g., direct break-up DNA molecules). The indirect action is more complex and depends heavily on the energy loss effects of radiation on the body tissue and subsequent chemistry. Both these effects are more susceptible to the biological targets. Dose modifying factor (DMF) is one of the most important parameters for the evaluation of radiation-induced damage. It is defined as the ratio of the dose to produce an effect in the presence of chemicals (radiomodifiers) to the dose to produce the same effect in the absence of chemicals. Radiosensitization occurs due to increase photo absorption of high *Z* number element resulting in the transfer of ionizing photon energy to the tumor tissue.

GNPs delivered to tumor tissues can selectively increase radiation efficiency leading to differentially increased cell killing. Theoretically the dose enhancement ratio of GNPs estimated [6, 7] using Monte carlo studies is significant (>200%). Bereco et al. reported tumor killing efficiency by GNPs with irradiation of 6 MV X-ray in a tumor of microvascular and endothelia cells [8]. Chitthrani et al. showed that 50 nm NPs radiosensitize in both lower and higher energy photon ranges with dose modification factor (DMF) of 1.66 for 105 kVp and 1.17 for 6 MvP [9, 10] reported radiation-induced enhancement factor of 25-fold Kv X-ray at 1 mM GNPs concentration in bovine endothelial cell with KV X-ray. GNPs enhancement radiation damage yield by factor greater than 2 using KV range and low-energy electrons in plasmid DNA model system [11, 12]. A comparison of GNP radiation therapy enhancement from literature survey is shown in Table 7.1.

Table 7.1 Comparison of GNP radiotherapy enhancement selected from the Literature

References	Paper citation	Cell lines	Radiation	GNP conc.	Effect
Jain et al. [34]	*Int. J. Radiat. Oncol.*, 2011, **79**, 531–539	MDA-MB-231	160 kVp	12 um	SER = 1.41
			6 MV	12 um	SER = 1.29
			15 MeV	12 um	SER = 1.16
			6 MeV electron	12 um	SER = 1.04
			16 MeV electrons	12 um	SER = 1.35
		DU 145	160 kVp	12 um	SER = 0.92
			6 MV	12 um	SER = 1.13
			6 MeV electron	12 um	SER = 1.12
		L132	160 kVp	12 um	SER = 1.05
			6 MV	12 um	SER = 1.08
			6 MeV electron	12 um	SER = 0.97
Chattpadhyay et al. [35]	*Breast Cancer Res. Treat.*, 2013, **137**, 81–91	MDA-MB-361	100 kVp	2.4 mg/ml	DEF = 1.3–1.6
Chang et al. [16]	*Cancer Sci.*, 2008, **99**, 1479–1478	B16F10	6 MeV electron	11 nm	DEF = 1.02
Chithrani [9]	*Radiat. Res.*, 2010, **173**, 719–728	HeLa	220 kVp	1 nm	DEF = 1.2–1.43
			105 kVp	1 nm	DEF = 1.66
			6 MV	1 nm	DEF = 1.17
			Cs-137 (662 keV)	1 nm	DEF = 1.18
Liu et al. [36]	*Phys. Med. Boil.*, 2010, **55**, 931–945	CT-26, EMT-7	160 kVp	0.4–1 mM	DEF = 1.1
		CT-26 EMT-8	6 MV	0.4–1.0 mM	DEF = 1
Daniel et al. [29]	*PLOS One*, 2013, **8**, e62425	U251	150 kVp	1 mM	SER = 1.3
Zhang et al. [37]	*Biomaterials*, 2012, **33**, 6408–6433, 6419	HeLa	Cs-137 (662 keV)	0.05 and 0.1 mM	DEF = 1.41–1. 65

Note: Enhancement can vary on the basis of several factors, including cell line and radiation source used.

GNP-induced radiosensitization in MV radiation energy range was studied by Dorsey et al. 2013 [13], who conducted studies on specialized radiochemical film to measure potential MV range energy dose enhancement. They suggested that GNPs are capable of radiosensitization in the clinically relevant MV range of radiation energy. GNP-induced radiosensitization due to the production of low-energy secondary electrons (which are about three times more efficient than X-rays), which causes DNA damage, operates at MV photon beam energies commonly used in radiotherapy [11].

The accumulation of GNPs in mammary carcinoma in mice tumor was shown by Heinfeld et al. [14]; when GNPs (1.9 nm) were injected via the intravenous route, they rapidly accumulated in tissues with 2.7 GNP body weight resulting in 7 mg GNPs/gm in the tumor immediately. Heinfeld et al. performed irradiation 60 s after injection with 250 kVp X-ray therapy. The significant survival was observed in the test system compared with the control experiments. In another report by Heinfeld, when mice bearing murine squamous cell sarcoma (SCC VII) were irradiated with X-rays (68 KeVp, 42 and 30 Gy), significant tumor growth delay and long-term tumor control were observed with 42 Gy but not with 30 Gy. Also, in mice irradiated with 157 KeV photon, more tumor radiosensitivity was seen with GNPs accompanied by 50.6 Gy than 44 Gy. Mice injected with melanoma cell (B16F10) were irradiated with electron (25 Gy) in the presence of GNPs; radiosensitization was observed in melanoma cells. Tumor growth rate decreased and the apoptotic signal increased [15]. Chang et al. showed that GNPs accumulate inside melanoma cells and enhance the efficiency of ionizing radiation, including tumor cell apoptosis and retarded tumor growth in tumor-bearing mice [16].

Proton radiotherapy increases tumor killing efficiency when directed against GNP-loaded tumors. Polf et al. 2011 [17] reported that the prostate tumor cell with internalized GPNs exhibited increase ionization density and lower surviving fraction when exposed to proton beams compared with cells exposed to proton therapy alone. Significantly increased DEF and reactive oxygen species (ROS) were observed in the mouse tumor loaded with GNPs and iron nanoparticles with proton energy (10–41 Gy), and increased tumor regression and mouse survival in vivo, too, were witnessed due to the released secondary electron and particle-induced radiation [18].

Butterworth [19] studied cellular cytotoxic response of cell lines to GNPs (1.9 nm) and measured the dose-modifying effect following exposure at low concentration. GNPs caused a significant level of cell type specific cytotoxicity, apoptosis, and increased oxidative stress. When dose-modifying agents were used, dose enhancement factor varied between the cell lines and showed the highest enhancement 1.9 in AO-1522B cells at GNP concentration of 100 ug ml^{-1}. Exposure to GNP (1.9 nm) to induce a range of cell lines–specific response including decrease clonogenic survival, increase apoptosis and induction of DNA damage.

7.3 GNP-Assisted Hyperthermia

Hyperthermia is the heating of tissue to a temperature ranging 40–45°C for a short period, which results in the destruction of cancer cells as they are heat sensitive and have low heat tolerance. Hyperthermia is a potential antitumor agent and has the ability to inhibit cell proliferation in human tumors. Hyperthermia can be carried out alone and in combination with ionizing radiation. Hyperthermia causes denaturation of enzymes, proteins, and nucleic acids; chromatin damage; cytoplasmic damage; membrane damage (permeability changes); and metabolic imbalance in cancer cells.

A number of clinical trials have studied hyperthermia in combination with radiotherapy and chemotherapy. These studies have focused on the treatment of several types of cancers such as sarcoma, melanoma, and cancers of the head and neck, brain, lung, esophagus, breast, and bladder. Many of these studies, but not all, have shown a significant reduction in the tumor volume when hyperthermia is combined with other treatments. However, not all of these studies have shown increased survival in the patient receiving the combined treatment. Treatment modalities of hyperthermia are generally local, regional, and whole-body hyperthermia. In local hyperthermia, heat is applied to a small area; different types of energy are used to apply heat, such as microwave, radiofrequency, and ultrasound. The external approach is used to tumors that are inside or just below the skin. External applications are positioned around the tumor to heat the appropriate region, and energies are focused on the tumor to raise its temperature. The intraluminal method may used to treat the tumors within or near body cavity such as esophagus. Probes are placed inside the cavity and inserted into the tumor site to deliver heat energy. The interstitial method is used to treat tumors deep within the body, such as brain tumors. In regional hyperthermia, different approaches are used to heat large areas of tissues such as body cavity, organ, and limb. Whole-body hyperthermia is used to treat metastatic cancers. This can be accompanied by several techniques that raise the body temperature (107–108°F), including the use of thermal chambers or hot water blanket.

The optothermal properties and biological targeting abilities of gold nanoparticles are much more promising in hyperthermic treatment. These features also make them a new-generation photothermal contrast agent for photothermal therapy. Another important feature of GNPs is enhanced permeability and retention factor (EPR), which enables its use in hyperthermia. The EPR effect is the property by which a chemical agent/drug accumulates in the tumor tissue more than the normal tissue. Pissuwana et al. [20] reported the EPR effect of gold that can be used in the local application of heat. The sources for heating such as the infrared lamp, lasers, and ultrasound can be used in the process; however, there is the limitation of heat generation in the targeted region. GNPs induce sufficient heat to increase cellular damage via thermal effects.

GNPs absorb light in the visible spectrum with peak absorption at 530 nm. Killing of cells using laser light in the presence of GNPs has been reported in vitro [21]. Spherical gold nanoparticles absorb only UV and visible light; so they are a poor choice of heating for tissue, since the penetration of UV visible light in the tissue is limited. The optimal wavelength for tissue penetration is near infrared, i.e., 800 nm. Solid spherical GNPs do not absorb significantly at 800 nm. Treatment of murine tumors [22] and subcutaneous tumors [23] with gold nanoshell and laser irradiation has been reported.

It is possible to achieve photothermal therapy by using spherical GNPs with pulse laser to the SPR absorption in the visible region for the treatment of skin cancers [24]. Lin and co-workers have performed hyperthermic treatment on targeted lymphocyte cells using laser and gold nanospheres [25]. Lymphocytes incubated with GNPs conjugated with antibodies exposed to nanosecond laser pulses showed cell death with 100 laser pulses at the energy of 0.5 J/cm^2. GNPs (40 nm) were conjugated the with anti-EGFR antibodies and targeted toward human neck cancer cells, detected by dark field light scattering and surface plasmon absorption spectra on single cells. The GNP-induced cancer cell damage at 19 w/cm^2 after the irradiation with Ar+ laser at 514 nm for 4 min, while heating cell do not show any loss of cell viability. The NIR photo thermal therapy both in vitro and in vivo was demonstrated by Hirsch et al (2003) using GNPs [26]. When breast carcinoma cell incubated with PEGylated gold nanoshells which unable absorption in the NIR region. Cells undergo irreversible photothermal damage after exposure to NIR light. The changes were observed coagulation, cell shrinking, loss of nuclear staining, tissue destruction, etc.

7.4 GNPs Targeted Therapy

Targeted drug therapy is an advanced drug therapy technique that delivers a certain amount of therapeutic agent for prolonged period to the targeted area within the body. This helps maintain the required plasma and tissue drug levels in the body. The delivery system is prepared by knowing specific features/properties of targeted cells, transport carriers, or vehicles which convey the drug to specified receptor ligands. The system must be biochemically inert, non-immunogenic, and biophysically stable. The targeted drug delivery system is generally categorized as passive and active targeting delivery systems. In passive targeting, the accumulation of the drug or drug carrier system takes place at specific sites. In an active targeting system, specific ligand receptors are involved in drug delivery.

Different functional groups can be attached to gold nanoparticles such as thiol, peptides, and antibodies to increase therapeutic index. Conjugation of GNPs (14 nm) with thioglucose increases uptake in ovarian cancer cell lines after 8–96 h [27]. Significant inhibition was observed relative to the control at 5–20 Gy. Cystamine and thioglucose coated with GNPs (15 nm) were studied in breast cancer cell lines [28]; cystamine-coated particles were taken up 3–4 times more efficiently than glucose-coated particles. Glucose-coated particles have increased radiosensitization than cystamine-coated particles. Larger-size GNPs (57 and 84 nm) were tested on HeLa cell lines. Equal uptake was noticed in both cases; however, unconjugated particles showed greater radiosensitization effect. In another report, PEGlyated and herceptin conjugated with 30 nm GNPs used for the study of uptake in MDA-MB 361 breast cancer tumor, where 11 Gy of 100 kVp image-guided X-ray irradiated was performed 24 h after injection, resulted in 46% reduction in tumor size compared with control experiment. No damage was noticed in normal cells. Folate was conjugated with gold nanorods (12.5 nm) and studied on MG 803 human gastric carcinoma cells, 60% decrease in cell viability was seen at 6 Gy X-ray irradiation relative to cells with gold nanorods.

A study by Daniel et al. [29] reported cell culture and animal model of GBM, where RT was performed along with PEGylated gold nanoparticles; GNPs significantly increased cellular DNA damage inflicted by IR in human GBM–derived cell lines and resulted in reduced clonogenic survival (with DER –1.3). Combined GNPs and RT also increased DNA damage to brain blood vessels.

Therapeutic effect, apoptosis, and tumor volume were investigated using Arg-Gly-Asp-conjugated GNPs integrated with radioiodine-125 on NCl-H446 tumor-bearing mice via Tc-99m-Annexin V SPECT. Significant radiosensitization was observed in treated cells compared with non-treated cells [30]. Glucose-capped GNPs enhanced radiosensitivity in human prostate cancer cells. Glucose capped GNPs (Glu-GNPs) triggered cell cycle acceleration in the G0/G1 phase and restrained in the G2/M phase. In case of HeLa cell lines [31] and ovarian cancer [32], enhanced radiosensitivity was observed with Glu-GNPs [33].

7.5 Conclusions

Remarkable physicochemical properties and customizable surface of gold nanoparticles make them an immensely potential candidate for cancer therapy. One of the important problems of nanoparticles is rapid excretion by the kidneys. The amount of gold required is more than 2 gm/kg of body weight, which is a large amount for human use. For achieving therapeutic levels using gold in tumor with less delivered total gold is needed.

Optimization of size, shape, surface property, and targets of gold may improve circulation time and accumulation in the tumor. Studies indicate that GNRT requires substantial further investigation. It is needed to optimize the conjugated targeting agents (one or two) with GNPs, which accelerate therapeutic index. More studies are needed to optimize GNPs surface architecture, mechanisms of action of killing tumor cells and also need to see the more safe clinical applications.

References

1. Kobayashi, H., Watanabe, R., Choyke, P. L. (2013). Improving conventional enhanced permeability and retention (EPR) effects; what is the appropriate target, *Theranostics,* **4**, pp. 81–89.
2. Mariam, J., Dongre, P. M., Kothari, D., Study of interaction of silver nanoparticles with bovine serum albumin using fluorescence spectroscopy, Flourec, J. (2011). DOI 10.1007/s10895-011-0922-3.
3. Mariam, J., Sivakami, S., Dongre, P. M. (2015). Albumin corona on nanoparticles—A strategic approach in drug delivery, *Drug Deliv.,* DOI: 10.3109/ 10717544.2015.1048488.
4. Christof, M. N. Nanoparticles, proteins and nucleic acids: biotechnology meets materials sciences, *Angew Chem.* (2001). Int Ed **40,** pp. 4128–4158.
5. Dejing, G., Tian, Y., and Shuyun, B. (2005). Studies on the interaction of colloidal gold and serum albumins by spectral methods, *Spectrochem. Acta Part A,* **62**, pp. 1203–1208.
6. Hainfeld, J. F., Dilmanian, F. A., Slatkin, D. N., et al. (2008). Radiotherapy enhancement with gold nanoparticles, *J. Pharm. Pharmacol.,* **60**, pp. 977–985.
7. Cho, S. H. (2005). Estimation of tumor dose enhancement due to gold nanoparticles during typical radiation treatments: A prelimeinary Monte Carlo study, *Phys. Med. Biol.,* **50**, N163–N173.
8. Berbeco, R. I., Ngwa, W., Makrigiorgos, G. M. (2011). Localized dose enhancement to tumor blood vessel endothelial cells via megavoltage X-rays and targeted gold nanoparticles: New potential for external beam radiotherapy, *Int. J. Radiat. Oncol. Biol. Phys.,* **81**, pp. 270–276.
9. Chithrani, D. B., Jelvesh, S., Jalali, F., et al. (2010). Gold nanoparticles as radiation sensitizers in cancer therapy, *Radiat. Res.,* **173**, pp. 719–728.

10. Rahman, W. N., Bishara, N., Ackerly, T., He, C. F., Packson, P., Wong, C., Davidson, R., G. (2009). Enhancement of radiation effects by gold nanoparticles for superficial radiation therapy, *Nanomed. Nanotechnology. Biol. Med. Nanomed.*, **5**, pp. 136–142.

11. Zhang, Y., Hunting, D. J., Ayotle, P., et al. (2008). Radiosensitization of DNA by gold nanoparticles irradiated with high-energy electrons, *Radiat. Res.*, **169**, pp. 19–27.

12. Foley, E. A., Carter, J. D., Shan, F., Guo, T. (2005). Enhanced relaxation of nanoparticle-bound supercoiled DNA in x-ray radiation, *Chem. Commun.*, **25**, pp. 3192–3194.

13. Jay, F., Dorsey, et al. (2013). Gold nanoparticles in radiation research: Potential applications for imaging and radio sensitization, *Trans. Cancer Res.*, **2**(4), pp. 280–291.

14. Hainfeld, J. F., Slatkin, D. N., Smilowitz, H. M., (2004). The use of gold nanoparticles to enhance radiotherapy in mice, *Phys. Med. Biol.*, **49**, pp. N309–N315.

15. Hainfeld, J. F., Dilmanian, F. A., Zhang, Z., et al. (2010). Gold nanoparticles enhance the radiation therapy of a murine squamous cell carcinoma, *Phys. Med. Biol.*, **55**, pp. 3045–3059.

16. Chang, M. Y., Shiau, A. L., Chen, Y. H., et al. (2008). Increased apoptotic potential and dose-enhancing effect of gold nanoparticles in combination with single-dose clinical electron beams on tumor-bearing mice, *Cancer Sci.*, **99**, pp. 1479–1484.

17. Polf, J. C., Bronk, L. F., Driessen, W. H., et al. (2011). Enhanced relative biological effectiveness of proton radiotherapy in tumor cells with internalized gold nanoparticles, *Appl. Phys. Lett.*, **57**, pp. 8309–8323.

18. Kim, J. K., Seo, S. J., Kim, H. T., et al. (2012). Enhanced proton treatment in mouse tumors through proton irradiated nanoradiator effects on metallic nanoparticles, *Phys. Med. Biol.*, **57**, pp. 8309–8323.

19. Butterworth, K. T., Coulter, J. A., Jain, S., et al. (2010). Evaluation of cytotoxicity and radiation enhancement using 1.9 nm gold particles: Potential application for cancer therapy, *Nanotechnology*, **21**(29), p. 295101.

20. Pissuwana, D., Valunzuelaa, S. M., Cortiea, M. B. (2006). Therapeutic possibilities of plasmonically heated gold nanoparticles, *Trends Biotechnol.*, **24**(2), pp. 62–67.

21. El-sayed, I. H., Huang, X., El-sayed, M. A. (2006). Selective laser photo-thermal therapy of epithelial carcinoma using anti-EGFR antibody conjugated gold nanoparticles, *Cancer Lett.*, **239**(1), pp. 129–135.

22. Von, M. G., Park, J. H., Agrawal, A., Bandaru, N. K., Das, S. K., Sailor, M. J., et al., (2009). Computationally guided photothermal tumor therapy using long-circulating gold nanorod antennas, *Cancer Res.*, **69**, pp. 3892–3900.

23. Neal, D. P., Hirsch, L. R., Halas, N. J., Payne, J. D., West, J. L. (2004). Photo-thermal tumor ablation in mice using near infrared-absorbing nanoparticles, *Cancer Lett.*, **209**, pp. 171–176.

24. Zharov, V. P., Galitovskaya, E. N., Johnson, C., Kelly, T. (2005). Synergistic enhancement of selective nanophotothermolysis with gold nanoclusters: Potential for cancer therapy, *Lasers Surg. Med.*, **37**(3), pp. 219–226.

25. Pitsillides, C. M., Joe, E. K., Wei, X., Anderson, R. R., Lin, C. P. (2003). Selective cell targeting with light-absorbing microparticles and nanoparticles, *Biophys. J.*, **84**(6), pp. 4023–4032.

26. Hirsch, L. R., Stafford, R. J., Bankson, J. A., Sershen, S. R., Rivera, B., Price, R. E., et al. (2003). Nanoshell-mediated near-infrared thermal therapy of tumors under magnetic resonance guidance, *Proc. Natl. Acad. Sci. (U.S.A.)*, **100**(23), pp. 13549–13554.

27. Geng, F., Song, K., Xing, J. Z., Yuvan, C. Z., Yan, S., Yang, Q. F., et al. (2011). Thio-glucose bound gold nanoparticles enhance radio-cytotoxic targeting of ovarian cancer, *Naotechnology*, **22**, p. 285101.

28. Kong, T., Zeng, J., Wang, X. P., Yang, X. Y., Yang, J. (2008). Enhancement of radiation cytotoxicity in breast-cancer cells by localized attachment of gold nanoparticles, *Small*, **4**, pp. 1537–1543.

29. Daniel, Y. J., Lova, S., Melissa, S., et al. (2013). Selective targeting of brain tumors with gold nanoparticle-induced radiosensitization, *PLOS ONE*, **8**(4), p. e62425.

30. Ning, S. U., Yajie, D., Guangli, L., Guizhi, L. (2015). Iodine-125-labeled cRGD-gold nanoparticles as tumor-targeted radiosensitizer and imaging agent, *Nanoscale Res. Lett.*, **10**, p. 160.

31. Kaura, H., Punjaria, G., Semwalb, M. K., Sarmma, A., Kumar, A. D. (2013). *In vitro* studies on radiosensitization effect of glucose capped gold nanoparticles in photon and ion irradiation of HeLa cells, *Nucl. Instr. Meth. Phys. Res.*, **301**, pp. 7–11.

32. Wang, C., Li, X., Wang, Y., Liu, Z. H., Fu, L. (2013). Enhancement of radiation effect and increase of apoptosis in lung cancer cells by thio-glucose bound gold nanoparticles at mega voltage radiation energies, *J. Nanoparticles Res.*, **15**, 1642.

33. Mohamad, B., Maryam, G. (2013). A systematic review of gold nanoparticles as novel therapeutics, *Nanomed. J.*, **1**(4), pp. 211–219.

34. Jain, S., Coulter, J. A., Hounsell, A. R., et al. (2011). Hirst DG, Cell-specific radiosensitization by gold nanoparticles at megavoltage radiation energies, *Int. J. Radiat. Oncol. Biol. Phys.,* **79**, pp. 531–539.

35. Chattopadhyay, N., Caiz, K. Y., et al. (2013). Molecularly targeted gold nanoparticles enhance the radiation response of breast cancer cells and tumor xenografts to X-radiation, *Breast Cancer Res. Treat.,* **137**, pp. 81–91.

36. Liu, C. J., Wang, C. H., Chen, S. T., et al. (2010). Enhancement of cell radiation sensitivity by pegylated gold nanoparticles, *Phys. Med. Biol.,* **55**, pp. 931–945.

37. Zhang, X. D., Wu, D., Shen, X., et al. (2012). Size-dependent radiosensitization of PEG-coated gold nanoparticles for cancer radiation therapy, *Biomaterials,* **33**, pp. 6408–6419.

Part 2

Effects of Ionizing Radiations on Biological Systems

Chapter 8

The Combined Effect of Hyper-Gravity and Gamma-Irradiation on Physiology of Wheat Seedlings

Sandhya Singh, Sagar S. Jagtap, and Pandit B. Vidyasagar

Physics Department, S.P. Pune University,
Ganeshkhind Road, Pune, Maharashtra, India

sandhya_c_singh@yahoo.co.in

In the past century, there has been a great rise in various space missions. In future there will be many manned space missions and considering our interest in space colonization, it is required that these space missions have a provision for various life support systems for crew members. Crop plants can provide food for crew members, revive the air by removing CO_2 and producing O_2, and purify water through transpiration. Hence, it is required to study the physiological response of various crops to altered-gravity and radiation environment, so as to design efficient bioregenerative life support systems.

8.1 Introduction

In order to sustain life in space, many experiments and space programs are being designed and implemented, so that we could understand the behavior of biological systems in space. In space

Radiation in Medicine and Biology
Edited by Pandit B. Vidyasagar, Sagar S. Jagtap, and Omprakash Yemul
Copyright © 2017 Pan Stanford Publishing Pte. Ltd.
ISBN 978-981-4745-92-5 (Hardcover), 978-1-315-20656-1 (eBook)
www.panstanford.com

farming, the key variables are gravity and radiations; hence, the study of the plant behavior in such conditions is required. There are many studies where the effects of hyper-gravity on plants [14, 16, 20–24] and the effect of gamma irradiation on dry seeds [1, 2, 4, 6–8, 15] were studied individually. As per our knowledge, very few studies have been carried out to study the combined effect of altered gravity and radiations on plants, but considering the conditions in space, it is mandatory to study the combined effect on biological systems. Also many experiments related to gamma irradiation effects on seeds carried out were dry before irradiation and then imbibed. However, Jun Kumagai [13] showed that the amount of radical does not decay for several months when irradiated seeds are kept dry at room temperature, it immediately decreases by absorbing water for one day at room temperature. Actual irradiation effect of a particular dose on seeds cannot be studied efficiently if they are irradiated and then imbibed since the water on imbibition may lower the effect. Consequently keeping all these factors in mind the current study was planned to study the combined effect of short-term hyper-gravity and gamma-irradiation on wheat seedlings, and the results obtained were compared to the results of effect of short-term hyper-gravity on wheat seedlings. Wheat (*Triticum aestivum* L.) being an important ingredient in our daily food intake was selected for the current study.

In the current study, initially the imbibed seeds were exposed to hyper-gravity stress (ranging from 300g to 1500g) and gamma-irradiation (dose ranging from 20 Gy–100 Gy) separately and after exposure these seeds were grown in normal conditions. The root length, shoot length, percentage seed germination, fluorescence parameters, and total proline content were observed on the 5th day of their growth. Later on, the imbibed seeds were exposed to both stresses (combined as hyper-gravity (300g–1500g) + gamma-irradiation dose of (40 Gy)) and same parameters were studied. The experiments were carried out on imbibed seeds because by exposing the imbibed seeds to short-term hyper-gravity or gamma-irradiation crucial information can be obtained, since several biochemical reactions are taking place at this stage and the seed is in a very active state of development.

8.2 Materials and Methods

Healthy wheat seeds (Lok-1 variety) were selected and surface sterilized by 0.5% fungicide and washed several times with distilled water before soaking them in distilled water for 24 h. After exposing these seeds to various treatments mentioned below, these seeds were sowed in 0.8% agar gel and growth parameters were studied on the 5th day of their growth. Statistical analysis such as significance *P*-value was calculated using Student's *T*-test, and standard deviation was also calculated.

8.2.1 Seed Selection

Wheat (*Triticum aestivum* L.) seeds of the variety Lok-1 were selected for the following studies. Lok-1 variety is grown mainly in the central zone of India. It has a high yield potential, great disease resistance, grain quality, and adaptability.

8.2.2 Combined Hyper-Gravity and Gamma-Irradiation Treatment

Pre-soaked seeds were placed in Eppendorf tubes filled with 1 ml (D/W) and exposed to hyper-gravity (300g, 600g, 900g, 1200g, 1500g) using a centrifuge (Super spin-R-V/FM, manufactured by Plasto Crafts Industries (P) Ltd, Mumbai, India) for 10 min. These seeds were then placed in a polythene bag and kept in a chamber of gamma radiation source (CO60) with dose rate of 3.95 Gy/min and irradiated to the dose 40 Gy. After the exposure these seeds were sown in 0.8% agar gel and kept in a plant growth chamber under controlled conditions (temperature 25°C, humidity 55 ± 10%, light intensity 18 μmol m^{-2} s^{-1}) for 5 days. Readings were taken on the 5th day of their growth.

8.2.3 Growth Parameters

Percentage seed germination, root and shoot length of the wheat seedlings were measured on 5th day. The root and shoot lengths were measured using a ruler with least count 1 mm.

8.2.4 Fluorescence Parameters

Leaves of 5-day-old wheat seedlings were dark adapted for 10 min, and the fluorescence parameters; the ratio Fv/Fm, which is a measurement that represents the maximum potential quantum efficiency of Photosystem II if all capable reaction centers were open; and the performance index (PI) of leaves, which is an indicator of sample vitality, were measured using a Handy PEA fluorimeter (Hansatech).

8.2.5 Proline Content

Free proline content in the leaves was determined by using the method described by Bates et al. [9]. Leaf samples (0.1 g) were homogenized in 2 mL of sulfosalycylic acid (3%) using mortar and pestle. About 1 mL of the extract was taken in a test tube and to it 1 mL of glacial acetic acid and 1 mL of ninhydrin reagent were added. The reaction mixture was boiled in a water bath at 100°C for 30 min. After cooling the reaction mixture, 2 mL toluene was added and then transferred to a separating funnel. After thorough mixing, the chromophore containing toluene was separated and absorption was read at 520 nm on a spectrophotometer. Toluene was used as blank. Concentration of proline was estimated by referring to standard curve of D-proline.

8.3 Results and Discussion

As compared to control there was a gradual decrease in the growth of the wheat seedlings with increase in the hyper-gravity stress. This can be clearly seen in Fig. 8.1.

Figure 8.1 Growth of control and treated wheat seedlings on 5th day.

8.3.1 Germination Percentage

Effect of hyper-gravity (300g–1500g) and gamma irradiation (40 Gy) on germination percentage was observed on the 5th day of growth of wheat seedlings. It was observed that the germination percentage decreases with increase in hyper-gravity value (as shown in Fig. 8.2); the highest percentage germination was observed in control and continuous decrease in it was observed for higher hyper-gravity values. The relative percentage decrease in germination percentage in wheat seedlings exposed to doses (300g + 40 Gy to 1500g + 40 Gy) as compared to the control were 16%, 34.6%, 37.64%, 61.89%, 76.32%, respectively. The relative percentage decrease in germination percentage in only hyper-gravity treatment was highest (38.43%) in 1500g (data not shown), which is very less than combined irradiation and hyper-gravity treatment.

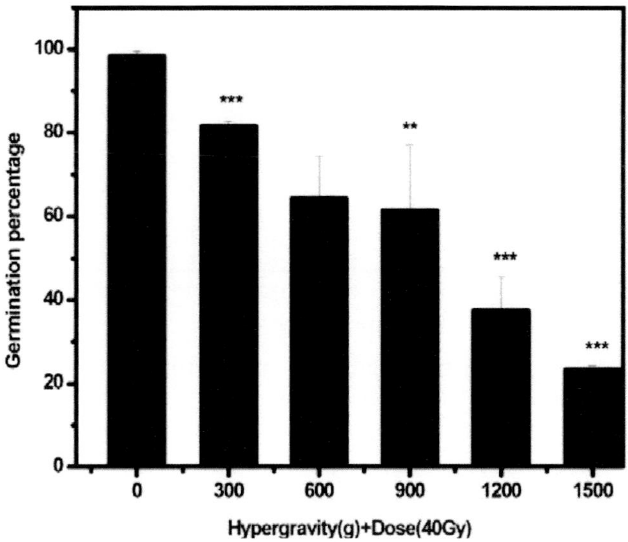

Figure 8.2 Germination percentage vs. hyper-gravity + gamma dose (40 Gy), (**) indicates $P < 0.05$ and (***) indicates $P < 0.001$.

8.3.2 Average Shoot Length

The effect of hyper-gravity (300g–1500g) and gamma irradiation (40 Gy) on the shoot length was observed on the 5th day of growth of wheat seedlings. The average shoot length showed a decreasing trend (as shown in Fig. 8.3), wherein the highest value of the shoot length was observed in the control and the lowest value of shoot length was observed in the seedlings that were exposed to 1500g + 40 Gy. The relative percentage decrease in the shoot length in wheat seedlings exposed to doses from (300g + 40 Gy to 1500g + 40 Gy) as compared to control were 17.79%, 38.48%, 49.88%, 59.42%, and 53.95%, respectively. The relative percentage decrease in shoot length in only hyper-gravity treatment was highest (65.37%) in 1500g (data not shown), which is comparable to the combined irradiation and hyper-gravity treatment.

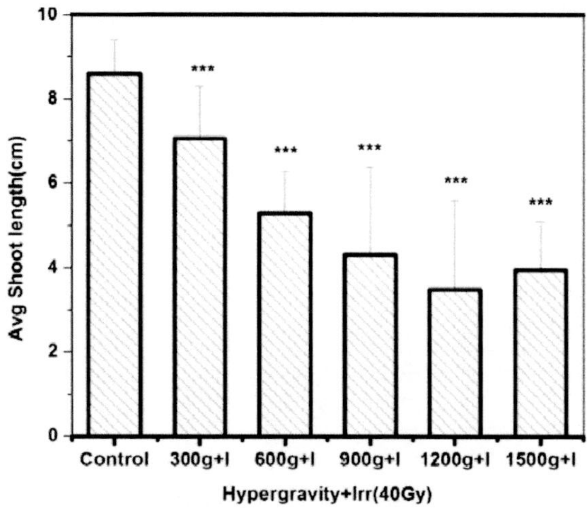

Figure 8.3 Average shoot length (cm) vs. hyper-gravity + gamma dose (40 Gy), (**) indicates $\boldsymbol{P} < 0.05$ and (***) indicates $\boldsymbol{P} < 0.001$.

8.3.3 Average Root Length

Effect of hyper-gravity (300g–1500g) and gamma irradiation (40 Gy) on the root length was observed on the 5th day of growth of wheat seedlings. The root length decreased with increase in hyper-gravity value (as shown in Fig. 8.4); the highest value of the root length was observed in the control and the lowest value of root length was observed in the seedlings that were exposed to 1500 + 40 Gy. The relative percentage decrease in the root length in wheat seedlings exposed to doses (300g + 40 Gy to 1500g + 40 Gy) as compared to control were 33.33%,

55.51%, 68.24%, 84.55%, and 82.69%, respectively. The relative percentage decrease in root length in only hyper-gravity treatment was the highest (66.98%) in 1500g (data not shown), which is very less compared to the combined irradiation and hyper-gravity treatment.

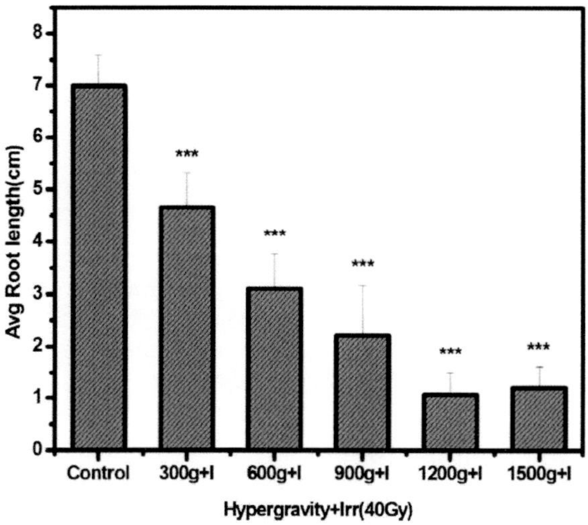

Figure 8.4 Average root length (cm) vs. hyper-gravity + gamma dose (40 Gy), (**) indicates $P < 0.05$ and (***) indicates $P < 0.001$.

8.3.4 Fluorescence Parameters

The effect of hyper-gravity (300g–1500g) and gamma irradiation (40 Gy) on the ratio Fv/Fm and the performance index (PI) of leaves was observed on the 5th day of growth of wheat seedlings. Progressive decrease in PI with increasing *g* value was observed. On the other hand, no significant decrease in Fv/Fm ratio values compared to the control was observed in all samples (as shown in Fig. 8.5a,b). The Fv/Fm ratio, which represents the maximum potential quantum efficiency of Photosystem II, did not differ much than the control, indicating that it may not be a sensitive parameter for stress [5]. On the other hand, the relative percentage decrease in performance index in wheat seedlings exposed to doses from (300g + 40 Gy to 1500g + 40 Gy) as compared to control was 16.88%, 23.37%, 26.62%, 25.97%, and 35.71%, respectively. The relative percentage decrease in performance index in only hyper-gravity treatment was highest (38.44%) in

1500g (data not shown) which is comparable to the combined irradiation and hyper-gravity treatment.

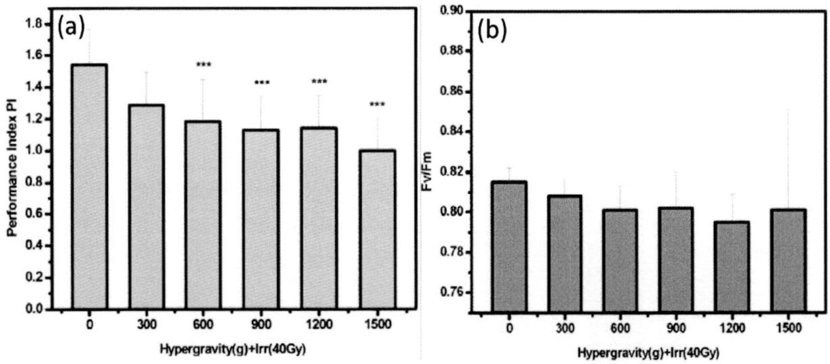

Figure 8.5 (a) Performance index vs. hyper-gravity + gamma dose (40 Gy), (**) indicates $P < 0.05$ and (***) indicates $P < 0.001$. (b) Fv/Fm vs. hyper gravity + gamma dose (40 Gy), (**) indicates $P < 0.05$ and (***) indicates $P < 0.001$.

8.3.5 Total Proline Content

The proline content extracted from leaves of the seedlings on 5th day of their growth and it was observed that the total proline content gradually increased with increase in the gamma dose

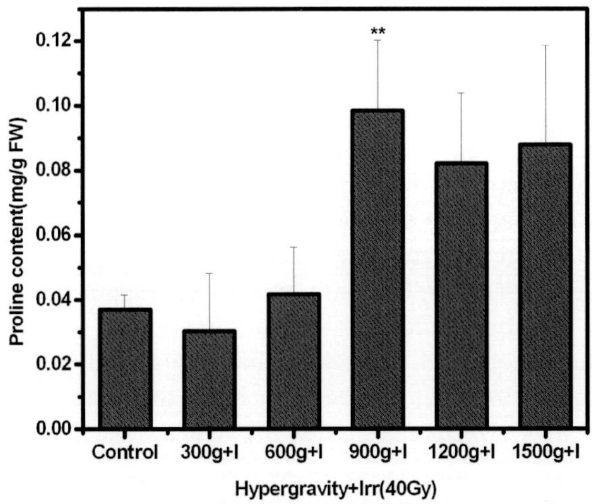

Figure 8.6 Proline content (mg/g FW) vs. hyper-gravity + gamma dose (40 Gy), (**) indicates $P < 0.05$ and (***) indicates $P < 0.001$.

wherein highest proline content was observed at 900g + 40 Gy dose and 300g + 40 Gy dose had lowest proline content (as shown in Fig. 8.6). The relative percentage decrease in proline content in 300g + 40 Gy compared to control was 18% and the relative percentage increase in proline content in (600g + 40 Gy–1500g + 40 Gy) doses compared to control were 9.7%, 62.24%, 54.87% and 57.47%, respectively. The relative percentage increase in proline content in only hyper-gravity treatment was the highest (84.13%) in 1200g (data not shown), whereas highest proline content was observed at 900g + 40 Gy (62.24%) in the combined hyper-gravity and gamma-irradiation treatment.

8.4 Conclusion

Previously, the effect of the gamma irradiation treatment (data not shown) on wheat seedlings and the effect of the short-term hyper-gravity treatment on wheat seedlings were studied (data not shown). Remarkable results were obtained in both studies, and this raised a question that if imbibed seeds were exposed to short-term hyper-gravity stress ranging from 300g to 1500g and each value was combined with the gamma irradiation dose of 40 Gy, what the resultant effect would be on the overall growth of the wheat seedlings, consequently inspiring the current study. The results of the combined effect of hyper-gravity stress ranging from 300g to 1500g and each stress value combined with the gamma irradiation dose of 40 Gy showed that it affects the overall growth, i.e., shoot length, root length, and also the photosynthetic process as it was observed that there was a gradual decrease in the performance index with increase in stress value, and it is known that the performance index is an indicator of sample vitality, thus indicating that the combined treatment also affects the photosynthetic process of wheat seedlings. It was also observed that in this case, the root length parameter was affected more than the shoot length parameter, which is similar to the results of the gamma irradiation treatment (data not shown); however, in this case the root length was affected to a greater extent compared to only the gamma irradiation treatment, this supports a statement commented by Kiefer and Pross [12] that the effect of combined interactions of irradiation and altered gravity on plants can be classified as additive, synergistic, or antagonistic. Thus, observing the results obtained in our study, it can be said that the combined effect of hyper-gravity and gamma irradiation has proved to be synergistic (increased the radiation effect). It was also observed that the proline content increased with increase in the stress value; however, it is noteworthy that even if proline is believed to help the plants overcome stress, no improvement was observed in the growth of the wheat seedlings. The possible reason for this might be the increased synthesis of proline and its decreased degradation under the stress [17]. Similar results were obtained by [3, 11], who studied proline content under salt and drought stress. Interestingly consistent increase in the proline content was only observed until 900g + 40 Gy, but the proline content deteriorated thereafter for higher stress values. One possible reason for this could be that initially only the degradation process of proline was affected, but increasing the stress value further inhibited the proline synthesis too. Therefore, further investigation of these results could possibly lead us to understanding the biosynthesis and degradation of proline, which may help us in understanding the role of proline in stress and its metabolic implications.

Thus, the current research was an attempt to contribute to the study of combined effect of altered-gravity and radiation on plants, which is more similar to the actual conditions in space. It is important to carry out such studies rather than studying the individual effects as it may not give us desired outcomes, since we have already seen that the combined effect was much more lethal to the growth of wheat seedlings and also its efficiency of overcoming the stress was decreased to a greater extent. Thus, it is essential to carry out such studies in order to plan the future space programs more proficiently.

Acknowledgments

The authors are grateful to the Board of College and University Development (BCUD), Savitribai Phule Pune University, Pune 411007, Maharashtra, India, and the Indian Space Research Organization (ISRO), India, for the financial support.

References

1. Dehpour, A. A., Gholampour, M., Rahdary, P., Jafari, M. R., Mohammad, S., and Hamdi, M. (2011). Effect of gamma irradiation and salt stress on germination, callus, protein and proline in rice (Oryza sativa L.) Determination of proline content. pp. 251–256.

2. Borzouei, A., Kafi, M., Khazaei, H., Naeriyan, B., and Majdabadi, A. (2010). Effects of gamma radiation on germination and physiological aspects of wheat. *Pak. J. Bot.*, **42**(4), pp. 2281–2290.

3. Delauney, A. J., and Verma, D. P. S. (1993). Proline biosynthesis and osmoregulation in plants. *Plant J.*, **4**, pp. 215–223.

4. Kiong, A. L. P., Lai, A. G., Hussein, S., and Harun, A. R. (2008). Physiological responses of Orthosiphon stamineus plantlets to gamma irradiation. *Am. Eurasian J. Sustain. Agric.*, **2**(2), pp. 135–149.

5. Stirbet, A., and Govindjee (2011). On the relation between the Kautsky effect (chlorophyll a fluorescence induction) and Photosystem II: Basics and applications of the OJIP fluorescence transient. *J. Photochem. Photobiol. B Biol.*, **104**(1–2), pp. 236–57.

6. Tabasum, A., Cheema, A. A. L. I., Hameed, A., Rashid, M., and Ashraf, M. (2011). Radio sensitivity of rice genotypes to gamma radiations based on seedling traits and physiological indices., **43**(2), pp. 1211–1222.

7. Baloch, A. W., Soomro, A. M., Bughio, H. R., Bughio, M. S., Mohammed, T., and Mastoi, N. (2004). Gamma irradiation induced chlorophyll mutations in rice (Oryza sativa L.). *J. Agric.*, **2**, pp. 257–261.

8. Akshatha, and Chandrashekar, K. R. (2013). Effect of gamma irradiation on germination growth and biochemical parameters of Pterocarpus santalinus, an endangered species of Eastern Ghats. **3**(2), pp. 266–270.

9. Bates, L. S., Waldren, R. P., and Teare L. D. (1973). Rapid determination of free proline for water stress studies. *Plant Soil*, **39**: 205–207.

10. Arnon, D. I. (1949). Copper enzyme polyphenoloxides in isolated chloroplast in Beta vulgaris. *Plant Physiol.*, **24**: 1–15.

11. Bohnert, H. J., and Jensen, R. G. (1996). Strategies for engineering water stress tolerance in plants. *Trends Biotechnol.*, **14**, pp. 89–97.

12. Kiefer, J., and Pross, H. D. (1999). Space radiation effects and microgravity. *Mutat. Res.*, **430**(2), pp. 299–305.

13. Kumagai, J., Katoh, H., Kumada, T., Tanaka, A., Tano, S., and Miyazaki, T. (2000). Strong resistance of Arabidopsis thaliana

and Raphanus sativus seeds for ionizing radiation as studied by ESR, ENDOR, ESE spectroscopy and germination measurement: Effect of long-lived and super-long-lived radicals. *Radiat. Phys. Chem.*, **57**, pp. 75–83.

14. Kozeko, L., and Kordyum, E. (2009). Effect of hypergravity on the level of heat shock proteins 70 and 90 in pea seedlings. *Microgravity Sci. Technol.*, **21**, pp. 175–178.
15. Rahimi, M. M., and Bahrani, A. (2011). Influence of gamma irradiation on some physiological characteristics and grain protein in wheat (*Triticum aestivum* L), **15**(5), pp. 654–659.
16. Searby, N. D., Steele, C. R., and Globus, R. K. (2005). Influence of increased mechanical loading by hypergravity on the microtubule cytoskeleton and prostaglandin E2 release in primary osteoblasts. *Am. J. Physiol. Cell Physiol.*, **289**(1), pp. C148–C158.
17. Kavi Kishor PB, Sangam S, Amruth RN, Sri Laxmi P, Naidu KR, Rao KRSS, Reddy KJSR, Theriappan P, Sreenivasulu N. Regulation of proline biosynthesis, degradation, uptake and transport in higher plants: its implications in plant growth and abiotic stress tolerance. *Curr. Sci.* 2005; **88**: 424–38.
18. Hare, P. D., and Cress, W. A. (1997). Metabolic implications of stress-induced proline accumulation in plants. *Plant Growth Regul.*, **21**(2), pp. 79–102.
19. Porra, R. J., Thompson, W. A., and Kriedmann, P. E. (1989). Determination of accurate extinction coefficients and simultaneous equation for assaying chlorophylls a and b extracted with four different solvents: Verification of the concentration of chlorophyll standards by atomic absorption spectroscopy. *Biochim. Biophys. Acta*, **975**, pp. 384–394.
20. Babbick, R., Dijkstra, M., Larkin, O. J., Anthony, P., Davey, M. R., Power, J. B., Lowe, K. C., Cogoli-Greuter, M., and Hampp, R. (2006). Expression of transcription factors after short-term exposure of Arabidopsis thaliana cell-cultures to hyper-gravity and simulated microgravity (2-D/3-D clinorotation, magnetic levitation). *Adv. Space Res.*, **39**, pp. 1182–1189.
21. Soga, K., Wakabayashi, K., Kamisaka, S., and Hoston, T. (2007). Effects of hypergravity on expression of XTH genes in azuki bean epicotyls. *Physiol. Plant*, **131**, pp. 332–340.
22. Yoshioka, T. R., Soga, K., Wakabayashi, K., Takeba, G., and Hoston, T. (2003). Hypergravity induced changes in gene expression in Arabidopsis hypocotyls. *Adv. Space Res.*, **31**, pp. 2187–2193.

23. Vidyasagar, P. B., Jagtap, S. S., Dixit, J. P., Kamble, S. M., and Dhepe, A. P. (2014). Effects of short-term hypergravity exposure on germination, growth and photosynthesis of *Triticum aestivum* L. *Microgravity Sci. Technol.*, **26**, 375–384. doi:10.1007/s12217-014-9400-2.

24. Vidyasagar, P., Jagtap, S., Nirhali, A., Bhaskaran, S., and Hase, V. (2008). Effects of hypergravity on the chlorophyll content and growth of root and shoot during development in rice plants. *Photosynthesis. Energy from the Sun: 14th International Congress on Photosynthesis* (pp. 1599–1602).

Chapter 9

The Study of the Effect of UV-C Radiation on the Current–Voltage Characteristics of Chitosan Membranes

Ni Nyoman Rupiasih,[a,b] Made Sumadiyasa,[a] and Putu Erika Winasri[a]

[a]*Department of Physics, Faculty of Mathematics and Natural Sciences, Udayana University, Bali, Indonesia*
[b]*Group Research of Material Sciences: Polymer and Biomaterial, Faculty of Mathematics and Natural Sciences, Udayana University, Bali, Indonesia*

rupiasih@gmail.com

This chapter presents a study in which the effect of UV-C radiation on the characteristics of the current–voltage (*I–V*) curve in a chitosan membrane was studied. The membrane used was chitosan membrane 2%. Irradiation was done at ambient temperature in air media. The membranes were irradiated for short time intervals, e.g., 10, 20, 40, and 60 min, and some of them were kept without irradiation, which were used as a control. The ion transport experiments were carried out using a cell membrane model that was composed of two compartments. The electrolyte solution used was NaCl with concentrations of 0.1 and 10 mM and the applied currents were 19–24 μA with increment of 0.5 μA. The potential difference was measured using a pair of Activon AEP calomel electrodes. All measurements were conducted

Radiation in Medicine and Biology
Edited by Pandit B. Vidyasagar, Sagar S. Jagtap, and Omprakash Yemul
Copyright © 2017 Pan Stanford Publishing Pte. Ltd.
ISBN 978-981-4745-92-5 (Hardcover), 978-1-315-20656-1 (eBook)
www.panstanford.com

at room temperature. The results obtained that the current–voltage curves of both groups, i.e., control and irradiated membranes, showed the ohmic behavior (region I). The increasing exposure time caused an increase in potential difference across the membrane, which led to a decrease in the membrane conductivity. It also decreased the water uptake capability of the membrane. The results indicate that a very short UV-C irradiation (10 min) was able to change the ionic transport properties and the water uptake of chitosan membrane, which increased with increase in time.

9.1 Introduction

Chitosan is a biopolymer derived by deacetylation of chitin. Chitin is an abundant biopolymer obtained mainly from shellfish such as shrimps, crabs, and prawn waste as a byproduct of the seafood industry. Due to the amine ($-NH_2$) and hydroxyl (–OH–) groups in its structure, which function as ligands, chitosan is a chelating polymer with excellent adsorption capacities for a number of metal ions [1–4]. Those groups also act as potential sites, where intermolecular interactions or modifications in the chemical functionality take place. Chitosan is also characterized by being cationic, biocompatible, and antibacterial. These characteristics of chitosan have been widely explored in different processes and applications, such as metal complexation, enzyme immobilization, controlled drug release, and sorption phenomena [4]. Chitosan membranes have been used for active transport of chloride ions in aqueous solution. The other applications are as a carrier and/or a selective barrier controlling the transport rate of the substances involved [2]. Chitosan is readily converted to fibers, membranes, coatings, powders, and solutions, further enhancing its usefulness [2, 5].

The numerous membranes have been developed for use in reverse osmosis, nanofiltration, ultrafiltration, microfiltration, pervaporation separation, and electrodialysis and in medical use such as artificial kidney [6]. Among these membranes, ion-exchange membranes are one of the advanced separation membranes. It has been used not only as electrodialysis concentration or desalting of solutions, diffusion dialysis to recover acids and electrolysis of sodium chloride solution but also in various fields as a polymeric film having ionic groups [7].

The ion-exchange membranes are classified into anion-exchange membranes (AEM) and cation-exchange membranes (CEM) depending on the type of ionic groups attached to the membrane matrix [6, 8, 9]. Cation-exchange membranes contain negatively charged groups fixed to the membrane backbone and allow the passage of cations but reject anions, whereas anion-exchange membranes contain positively charged groups fixed to the membrane backbone and allow the passage of anions but reject cations.

A method frequently used to characterize the transport properties of ion-exchange membranes is to study the *I*–*V* curves corresponding to the membrane system. The current–voltage curve usually reflects the electric property of the membrane and provided information about the transport mechanism of ions, including concentration polarization [9, 10]. When the electric current passes through a membrane system, the current flows in the solution through cations and anions, whereas in the membrane current flows mainly by counterions. The differences between the mobility of the counterion in the membrane and that in the solution phase lead to a depletion in the electrolyte concentration on the membrane surface. As a consequence, a concentration gradient develops in the thin film that adjoins the membrane surfaces, called a concentration polarization phenomenon. The current–voltage curve of an ion-exchange membrane has a characteristic shape, which shows three regions (Fig. 9.1) [9, 10]. The first region (Region I) is the ohmic region, where the current density is related to the electrical potential difference ruled by Ohm's law. The resistance of the system could be approximately attributed to ionic transport into the ion-exchange membrane (R_{ohm}). This region observed at low current density. The second region (Region II) is limiting current region, where current density varies very slowly with the potential to form a pseudo-plateau. This is the region of the limiting current density (LCD). The limiting current density is the current needed to transfer all the existing ions. In accordance with the concentration polarization theory, the limiting current value can give information about the thickness of the diffusion boundary layer, the diffusion coefficient, or the cation transport number in the membrane. The third region (Region III) is the electroconvection region, where the slope of the current–

potential curve increases again (R_{ec}); R_{ec} is the R value of the third region, which is usually lower than that in region I. According to the classical theory of concentration polarization, currents larger than the LCD may not be expected but in practice, overlimiting currents (third region III) can exist. The shape of the current–voltage curve in an ion-exchange membrane varies with the external conditions, such as concentration, flow rate, and the physicochemical conditions of the membrane surface [9, 10].

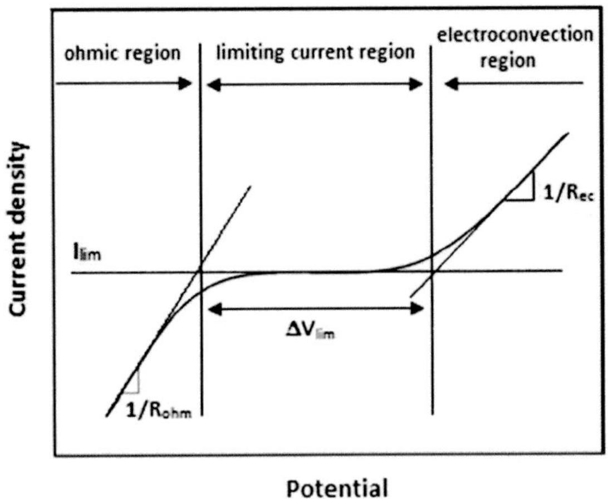

Figure 9.1 Typical current–voltage (I–V) curve of a cation-exchange membrane showing the ohmic region (Region I), limiting current regions (Region II) and an overlimiting current (Region III) [9, 10].

Some studies about the structural modifications in chitosan-based systems induced by UV irradiation have been reported [4]. In chitosan films exposed to low-intensity UV lamps, measurements of the static contact angle have revealed an enhancement in the surface energy with the exposure time [11]. This behavior has been related with the increase of the surface polarity of films caused by the scission of glycosidic bonds and pyranose rings during the photo-oxidation process [12, 13]. UV effects have also been studied on polymeric blends containing chitosan, where a significant reduction of UV sensitivity has been observed. However, chitosan-based blends show a lowering of the mechanical properties upon UV irradiation when compared with pure chitosan films [14–16].

Although the effects of UV irradiation on the physical properties of pure chitosan films and chitosan-based polymeric blends have been widely investigated, the effects of UV-C radiation on the ion transport properties of chitosan membrane have not been explored yet. In the present work, we study the *I–V* curves characteristic of chitosan membranes. The water uptake properties of them were also reported. The current–voltage curves for different concentrations of electrolyte solution were obtained under the same experimental conditions.

9.2 Materials and Methods

9.2.1 Materials

Chitosan (shrimp-based product) used are in the form of powder with properties of degree of deacetylation 87.4%, solubility in acetic acid is 99.4%, viscosity of 663,000 cps and average molecular weight of 900,000 g/mol. Acetic acid and sodium hydroxide p.a. were analytical grade and were used without further purification.

9.2.2 Preparation of Chitosan Membranes

The chitosan membranes were prepared by the evaporation of solvent from the casting solution. The preparation consisted of dissolving chitosan in acetic acid 1% (v/v) solution with 2% wt chitosan concentration. The solution was filtered in order to remove undissolved chitosan and debris and cast on a glass plate sized of 18.5 cm × 24.0 cm. After drying at room temperature for 6 days, the membranes were immersed in 1 M NaOH for 12 min for neutralization. They were then washed intensively with distilled water to completely remove NaOH and dried at room temperature again. Finally, it obtained dry chitosan membranes, which are ready to characterize or use.

9.2.3 Current–Voltage Measurement

The ion transport experiments were carried out in a cell membrane model composed of two compartments, and the applied currents were 19–24 μA with increment of 0.5 μA. The electrolyte solution used was NaCl with concentrations of 0.1

and 10 mM. The membrane was placed in a circular hole between the compartments, Fig. 9.2. Both compartments were filled with equal volume (13 mL) of solution. The effective area of the membrane was 1.77 cm^2. The potential difference across the membrane was measured by using two calomel electrodes.

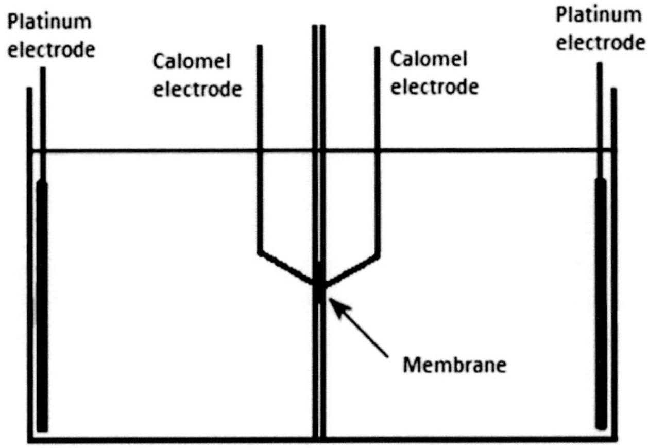

Figure 9.2 Schematic diagram of a two-compartment cell used in current–voltage measurements. Two platinum electrodes are connected to potentio/galvanostat and two calomel electrodes to the multimeter to measure the membrane potential.

The electrical current was supplied by a potentiostat connected by one pair of platinum electrodes. The membranes were irradiated for 10, 20, 40, and 60 min, and a few of them were kept without irradiation, which were used as a control. Further, the control and irradiated membranes were named M0, M10, M20, M40, and M60, respectively. All the measurements were conducted at room temperature of ± 28.8°C.

9.2.4 Water Uptake by the Membranes

The sample membranes were cut in 1 cm^2 size and dried in vacuum at room temperature for 6 days. The dry weight of the membrane (W_{dry}) was immediately measured. After dry weighing, they were kept in a beaker with 50 mL of distilled water at room temperature for 24 h. Then the membrane was taken out and the excess water was removed from the surface of the membrane with a filter paper. The weight of a wet membrane (W_{wet}) was measured. The water uptake by the membrane was calculated using the following formula [3, 17, 18]:

$$\text{Water uptake (\%)} = \frac{W_{wet} - W_{dry}}{W_{dry}} \times 100\% \qquad (9.1)$$

9.3 Results and Discussion

9.3.1 Characteristics of Current–Voltage Curves

Figures 9.3 and 9.4 show the *I*–*V* curves obtained for the chitosan membranes control (M0) and with UV-C irradiation, M10, M20, M40, and M60, in contact with NaCl solution with concentrations of C1:C2 of 0.1 mM:0.1 mM and 10 mM:0.1 mM, respectively. Figure 9.3 shows that the current–voltage curve of each chitosan membrane is linear and ohmic (region I), where the current density is related to the electrical potential difference governed by Ohm's law. The higher the applied current to the system, the higher the potential difference (voltage) exist at the membrane. This means the resistance of system (R_{ohm}) could be attributed to ionic transport in the anion-exchange membrane. In our study, the current–voltage curves obtained did not show regions II and III, the typical the current–voltage curves of an ion-exchange membrane, which can be divided into three regions as shown in Fig. 9.2.

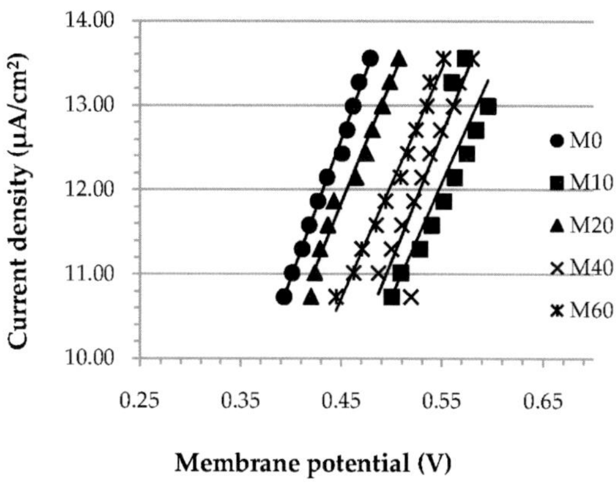

Figure 9.3 The current–voltage curves of chitosan membranes, M0, M10, M20, M40, and M60, in contact with NaCl solution with concentrations of C1 and C2 of 0.1 mM:0.1 mM.

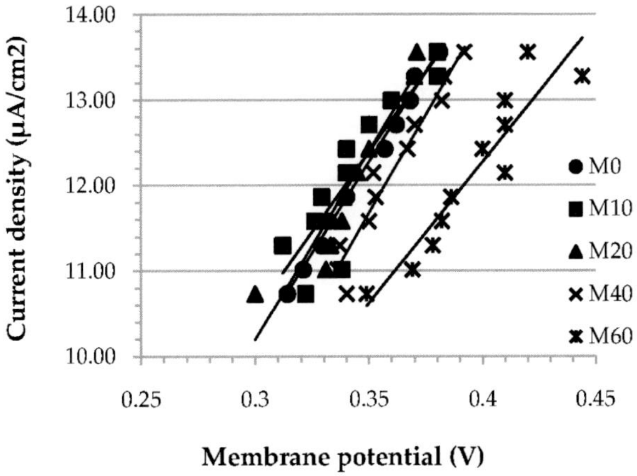

Figure 9.4 The current–voltage curve of chitosan membranes, M0, M10, M20, M40, and M60, in contact with NaCl solution with difference concentrations of C1 and C2 of 10 mM:0.1 mM.

The similar patterns obtained in Fig. 9.4, the current–voltage curves of each chitosan membrane is linear and ohmic (region I). However, the distribution of the data is fluctuating compared with the data presented in Fig. 9.3, especially for those membranes that have been irradiated with UV-C. This might indicated that some defects have occurred on the irradiated membrane.

9.4 Conductance of Membranes

Figure 9.5 shows the conductance of membranes in both environments, i.e., in NaCl solution with concentrations of C1 and C2 (C1:C2) of 0.1 mM:0.1 mM and 10 mM:0.1 mM, respectively. It shows that UV-C radiation has lowered the conductance of chitosan membranes, which decreases its ability to transport ions in an electrolyte solution. An exception on M40 has been obtained in a solution of NaCl with C1:C2 of 10 mM:0.1 mM, where conductance increased. The figure also shows that the conductance of the membranes in the NaCl solution with concentrations of 10 mM:0.1 mM is greater than that in 0.1 mM:0.1 mM. This result shows that the difference of concentrations of the solution in both compartments acted as other driving forces, which increased the driving force in the ionic transport processes.

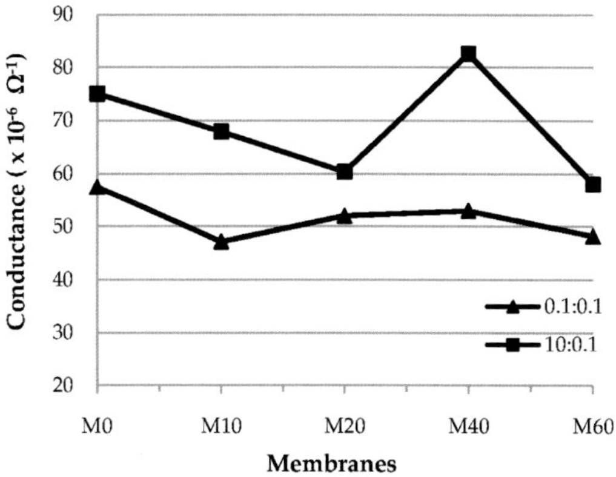

Figure 9.5 The conductance of chitosan membranes, M0, M10, M20, M40, and M60, in contact with NaCl solution with concentrations of C1 and C2 of 0.1 mM:0.1 mM and 10 mM:0.1 mM.

9.4.1 Water Uptake by Chitosan Membranes

Biomaterials are usually used in an aqueous environment; so the water uptake capability is an important property of biomaterials. Figure 9.6 shows the water uptake capability of all chitosan membranes, control (M0) and UV-C irradiated: M10, M20, M40, and M60. It was found that the water uptake capability of the membranes decreases with increasing time of UV-C exposure from 0 to 60 min. These results clearly indicate that UV-C radiation greatly lowers the water uptake capability of the membranes and they tend to be hydrophobic.

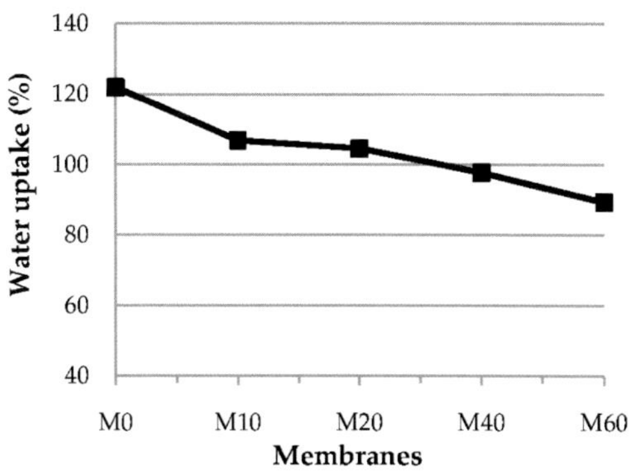

Figure 9.6 The percentage of water uptake of chitosan membranes, M0, M10, M20, M40, and M60.

9.5 Conclusion

The effect of UV-C radiation on the ionic transport properties of chitosan membrane has been observed from *I–V* measurements. The results clearly showed that the current–voltage curves of both groups of membranes, i.e., control (M0) and irradiated (M10, M20, M40, and M60) showed the ohmic behavior (region I). Increasing the time of UV-C exposure from 0 to 60 min caused a decrease in conductivity and the water uptake capability of the membranes. The study demonstrated that the ionic transport and the water uptake capability of chitosan membranes were affected by UV-C irradiation within very short time of exposure (i.e., 10 min). This observation suggested that more care should be taken during the sterilization processes and outdoor use of the membrane.

References

1. M. E. I. Badawy and E. I. Rabea (2011). A biopolymer chitosan and its derivatives as promising antimicrobial agents against plant pathogens and their applications in crop protection, *Int. J. Carbohydr. Chem.*, **2011**, 29 p.
2. N. N. Rupiasih, Y. E. Puspita, and M. Sumadiyasa (2015). Transport phenomena in chitosan synthetic membranes with emphasis on the effect of variations in the ratio of matrix/solvent, *J. Phys. Conf. Ser.*, **622**, 012004.
3. S. Sachdeva, R. P. Ram, J. K. Singh, and A. Kumar (2008). Synthesis of anion exchange polystyrene membranes for the electrolysis of sodium chloride, *Am. Inst. Chem. Eng. AIChE J.*, **54**(4), 940–949.
4. A. P. P. Praxedes, A. J. C. da Silva, R. C. da Silva, R. P. A. Lima, J. Tonholo, A. S. Ribeiro, and I. N. de Oliveira (2012). Effects of UV irradiation on the wettability of chitosan films containing dansyl derivatives, *J. Colloid Interface Sci.*, **376**, 255–261.
5. Y. Wan, K. A. M. Creber, B. Peppley, and V. T. Bui (2003). Synthesis, characterization and ionic conductive properties of phosphorylated chitosan membranes, *Macromol. Chem. Phys.*, **204**(5/6), 850–858.
6. M. Y. Kariduraganavar, R. K. Nagarale, A. A. Kittur, and S. S. Kulkarni (2006). Ion-exchange membranes: Preparative methods for electrodialysis and fuel cell applications, *Desalination,* **197**, 225–246.

7. T. Sata (1991). Ion exchange membranes and separation processes with chemical reactions, *J. Appl. Electrochem.*, **21**, 283–294.

8. T. Xu (2005). Ion exchange membranes: State of their development and perspective, *J. Membr. Sci.*, **263**, 1–29.

9. J.-H. Choi, H.-J. Lee, and S.-H. Moon (2001). Effects of electrolytes on the transport phenomena in a cation-exchange membrane, *J. Colloid Interface Sci.*, **238**, 188–195.

10. G. Chamoulaud and D. Bélanger (2005). Modification of ion-exchange membrane used for separation of protons and metallic cations and characterization of the membrane by current–voltage curves, *J. Colloid Interface Sci.*, **281**, 179–187.

11. A. Sionkowska, H. Kaczmarek, M. Wisniewski, J. Skopinska, S. Lazare, and V. Tokarev (2006). The influence of UV irradiation on the surface of chitosan films, *Surf. Sci.*, **600**(18), 3775–3779.

12. A. L. Andrady, A. Torikai, and T. Kobatake (1996). Spectral sensitivity of chitosan photodegradation, *J. Appl. Polym. Sci.*, **62**(9), 1465–1471.

13. N. Nyoman Rupiasih and P. B. Vidyasagar (2008). Comparative study of effect of low and medium dose rate of g irradiation on microporous polysulfone membrane using spectroscopic and imaging techniques, *Polym. Degrad. Stabil*, **93**, 1300–1307.

14. A. Sionkowska, M. Wisniewski, J. Skopinska, S. Vicini, and E. Marsano (2005). The influence of UV irradiation on the mechanical properties of chitosan/poly(vinyl pyrrolidone) blends, *Polym. Degrad. Stabil.*, **88**(2), 261–267.

15. A. Sionkowska, M. Wisniewski, J. Skopinska, G. F. Poggi, E. Marsano, C. A. Maxwell, and T. J. Wess (2006). Thermal and mechanical properties of UV irradiated collagen/chitosan thin films, *Polym. Degrad. Stabil.*, **91(**12), 3026–3032.

16. A. Sionkowska, J. Skopinska-Wisniewska, A. Planecka, and J. Kozlowska (2010). The influence of UV irradiation on the properties of chitosan films containing keratin, *Polym. Degrad. Stabil.*, **95**(12), 2486–2491.

17. P.-H. Chen, Y.-H. Hwang, T.-Y. Kuo, F.-H. Liu, J.-Y. Lai, and H.-J. Hsieh (2007). Improvement in the properties of chitosan membranes using natural organic acid solutions as solvents for chitosan dissolution, *J. Med. Biol. Eng.*, **27**(1), 23–28.

18. M. C. F. C. Felinto, D. F. Parra, C. C. da Silva, J. Angerami, M. J. A. Oliveira, and A. B. Lugão (2007). The swelling behavior of chitosan hydrogels membranes obtained by UV- and g-radiation, *Nucl. Instrum. Methods Phys. Res. B*, **265**, 418–424.

Chapter 10

Investigating Effects of Radiation Due to Cell Phones on Health Parameters of Youngsters during Continuous Conversation

Sakharam D. Aghav[a] **and Pandit B. Vidyasagar**[b]

[a]*Baburaoji Gholap College, Sangvi, Pune 27 (S.P.P.U, Pune), Maharashtra, India*
[b]*S.R.T.M University, Nanded, Maharashtra, India*

sdaghav@gmail.com

The continually increasing intensity of the invisible electromagnetic radiation due to cell phones affects human life in many ways. Base transmission towers erected for the better coverage of signals emit radiation frequently, which has unfavorable outcomes. Especially the younger generation is more and more addicted to phones and use them while talking, listening to music, downloading, playing games, and using various software applications.

In the present study, 50 healthy and young students were selected to observe variation in health parameters. These observations were noted using Multipara Caddo 19 B instrument, when the volunteers were exposed to radiation emitted by the

Radiation in Medicine and Biology
Edited by Pandit B. Vidyasagar, Sagar S. Jagtap, and Omprakash Yemul
Copyright © 2017 Pan Stanford Publishing Pte. Ltd.
ISBN 978-981-4745-92-5 (Hardcover), 978-1-315-20656-1 (eBook)
www.panstanford.com

cell phone when the volunteers communicated constantly with their friends. These observations were compared with the control. The changes in blood pressure, respiration rate, and percentage of saturated oxygen in the blood were not statistically significant even after mobile communication for 20 min. However, changes in the heart rate after 10 min exposure were at the threshold of significance (p = 0.0544) and became strongly significant (p = 0.0027) after mobile communication for 20 min.

10.1 Introduction

Investigations concerning electromagnetic energy can be divided into three major categories based on the sources. The first category is includes electromagnetic fields produced by electricity circulation system, electric circuits used in computers, television, microwave ovens, cell phones, and wireless communication towers systems. These fields are can be called techno fields. Moreover, human beings are facing electromagnetic radiation from natural sources such as the sun and the rest of the universe. These fields are called geofields. They exist in nature. They are unpredictable due to their characteristics of fluctuations such as eruption during solar eclipse. They are cyclic in nature. They are as due to diurnal and seasonal changes. Also, they are relatively stable like the earth's magnetic field. The ionosphere reflects all the radiation passing through it, including radio waves, which play a vital role in the communication. The third field is related to the human body. It is the field created by all living beings during metabolic activity in the body and, hence, is called biofields. The electricity produced within body—bioelectricity–is a biofield. It is due to the concentration gradient generated across the membrane of the cell.

There are number of devices that have become an essential part of our lives. Modern communications devices, medical devices, medical equipment, and electrical devices are commonly used devices. It is not possible to imagine life without them. The working of these devices is based on the principles of electricity and magnetism. Many devices employ electromagnetic radiation for their functioning. All living things on the earth ultimately depend on the electromagnetic radiation received from the sun. The main source of energy is the sun. We use various forms of the solar energy. Wind energy also depends on solar radiation. Solar energy is utilized by plants and animal to prepare their food.

The conversion of solar energy to food by photosynthesis in plants—the basic step in the food chain—depends on electromagnetic radiation. The process begins with the absorption of solar energy by photosynthetic pigments such chlorophyll a, chlorophyll b, and caretonoid. These pigments absorb photons in the radiation and initiate the food chain. Green plants are receptive to the solar radiation as they contain chlorophylls in large quantities. The light in form of radiation is absorbed by the chlorophyll pigment in the leaves and produces foodstuff necessary for the growth of a plant.

Solar spectrum comprises visible portion of a wide range of frequencies. Basically, all the energy reservoirs in the universe are stored forms of the energy received from the sun for millions of years. Electromagnetic radiations contain various energies in the form of waves traversing through space with the velocity of 3×10^8 m/s. They include radio waves, microwaves, visible, ultraviolet rays, and gamma rays. The frequency of waves differentiates these classes of radiation.

Our life is dependent on non-natural electromagnetic radiation (EMR), too. Following are a few examples of the applications of such electromagnetic radiation:

- warming of food a microwave oven
- use of radar waves to guide aircraft; radar waves are reflected from the aircraft and again received by the receiver in the radar system
- broadcasting of all radio stations and TV stations
- automatic self-focusing cameras
- cell phone communications
- radiation emitted by cell towers
- radiation generated by medical devices in radio therapy
- radiation generated in imaging techniques such as X-ray radiography, computer tomography, etc.
- radiation emitted by stray light
- radiation due to electricity power lines

- various electrical appliances such as refrigerator, hot plate, heater radio, computers, laptops, and Wi-Fi

Ultraviolet radiation cannot be seen by eyes, but its effect can be harmful. The same is the case of X-rays, which play an important role in diagnosis and treatment in medicine. X-rays allow physicians to examine the inner organs of the body, and any disorder in the functioning can be interpreted. The exposure to X-rays is kept to a minimum to avoid hazards of the invisible and high-energy radiation.

Electromagnetic radiation has a noticeable effect on living things on the earth and in the space. A variety of devices are used in communication, medicine, and applications based on electricity. Due to diverse applications, they become vital parts of the daily life. We are becoming more and more dependent on the communications technology. The globe is going to be flooded with electromagnetic radiation in the frequency range 50 Hz to 2 GHz. The technofield category includes electromagnetic fields produced by the electricity circulation system, electric circuits used in computers, television, microwave ovens, cell phones, and wireless communication towers systems.

Electromagnetic radiations can be classified into ionizing and non-ionizing radiation. Ionizing radiation has the capability to ionize and break the covalent bonds in atoms. Electromagnetic radiation has sufficient energy to remove the electrons in the outermost orbit of an atom, thereby creating charged particles or ions. All other forms of electromagnetic radiation are non-ionizing. The power of non-ionizing radiation is not sufficient to start the ionization process. However, the effects of non-ionizing radiation are substantial when radiations are irradiated for a longer time. It happen due to radiation power only. The radiation absorbed by the cell within the body produces heat in the body. These effects are called thermal effects. In these thermal effects, the radiation frequency is considerable. This energy is responsible for its dissemination into the organism. The continuous exposure to electromagnetic fields due to sources like power line appliances and sources of radiofrequency radiation has been linked with many adverse health effects [1, 2]. The types of non-ionizing radiation and the frequencies in which they occur are shown in Table 10.1.

Table 10.1 Non-ionizing radiation

Type	Frequency
Electromagnetic fields from direct current (static)	0 Hz
Waves from electric power (low-frequency)	50–60 Hz
Extremely low frequency (ELF) and very low frequency (VLF) fields	Up to 30 KHz
Radio frequencies (RF) including low frequency (LF), medium frequency (MF), high frequency (HF), very high frequency (VHF), ultra high frequency (UHF) and microwave (MW) and millimeter wave	From 30 kHz to 300 GHz
Infrared (IR) light, visible light and ultraviolet (UV) light	Above 300 GHz

Currently, communication devices such as wireless phones and cell phones are commonly used. The younger generation is more and more addicted to cell phones and use it frequently for communicating, transferring information via the Internet, watching videos and movies, listening to music, downloading and playing games, and various applications. They have all the information about newly introduced applications of cell phones and presume cell phone as a status symbol without knowing its ill effects. Electromagnetic or radiofrequency radiations are widely employed in communication devices. According to recently published BioInitiative report, there was abundant information about increase in danger and warnings about cancer, especially with regard to breast cancer in women. These reports also indicate that children and youngster are at a higher risk [3].

Presently, at global level there are more than 7 billion cell phone subscribers. In India, the cellular phone service was initiated on August 15, 1995, in Delhi on the non-commercial basis. In the past 20 years, the scenario has completely changed. There has been an exponential growth of mobile telephony in India and all over the world. As per data available from the Telephone Regulatory Authority of India, on April 15, 2015, 960,579,472 people were using cell phones all over India. The data indicate that 78% of India's population is using cell phones. To satisfy the demand, more than 20 companies provide mobile services in India and about 100 companies are manufacturing cell phones all over the world.

The number of companies manufacturing cell phones with various attractive features is continuously increasing every day. To provide better signal coverage for cell phones and connectivity, more than 70 million towers have been erected in India. The health hazards of cell phone towers are a topic of debate. Not a single company thinks about health hazards because of the business point of view. As not all of them living near tower are earning rents on towers: People staying near cell phone towers are not taking tower erection seriously. The increasing intensity of invisible and dangerous electromagnetic radiation from communication devices may affect life in numerous ways. In addition, base transmission towers emit electromagnetic radiation constantly, which has adverse effects on human as well as on wild life. Unfortunately, essential regulatory policies were not implemented considering the mammoth growth of mobile telephony.

There are reports of changes in the heart rate, and some cardiovascular effects with change in strength of immunity, including augmented sensitive and inflammatory responses, have been reported. In some studies, people living in proximity of very low intensity radiofrequency areas (like Wi-Fi and cell phone towers), were victims of sleep disturbance. Temporary effects on cognition, remembrance and learning, misconduct, response time, awareness and attentiveness and distorted brain wave activity were also observed [3].

Radiation emitted by cell phones may cause an increase in the permeability of the blood–brain barrier, which may result in neuronal damage and shows significant effects on genotoxicity and discharge of the blood–brain barrier also. In case of animals, the blood–brain barrier can be influenced by even a very low intensity radiofrequency exposure. There were reports of continuous 2 h exposures of cell phone radiation resulting in an increase in leakage of the barrier. Cell phone tower radiations can cause both thermal and non-thermal injuries. The damage can be caused due to heating at the tissue level, showing denaturation of proteins, disorder of molecule, changes in chemical reactions in the cell, heat shock proteins and even enzyme modifications in cells [4].

The genotoxic effects from radiofrequency exposure, such as chromosomal unsteadiness, distorted gene expression, and genetic, changes were reported [5]. A study regarding the effects on fertility and reproductive system revealed that there was a considerable reduction in the percentage of motile sperm and reproductive capacity [6, 7].

Difficulties in concentration, tiredness, sleep disorder, giddiness, loss of concentration, loss of memory and retentiveness, etc., were reported in case of electromagnetic hypersensitive (EHS) people. Reports of EHS people showed considerable variations in parameters of the autonomic nervous system (ANS), with changes in the capillary blood stream (microcirculation), heart rate, and electric skin potentials [8]. Scientists conducted a study to examine the association between cell phone use and possibility of glioma linked with cancer. Among the observations, several studies did not find any concrete proof of the augmented risk of glioma associated with the common use of cell phone. On the other hand, in case of people using cell phone for more than 10 years, the occurrence of tumor on the surface of the head was observed. The major probable danger was due to most heavily exposed part of the brain with long-term exposure of EMR [9, 10].

The study of electromagnetic field exposure has been conducted and found related with Alzheimer's syndrome, motor neuron disease, and Parkinson's disease [11]. It was observed that there was noticeable disparity between the heart rate (HR) during conversation on cell phone as compared with HR during cell phone kept in the ringing-and-resting mode in case of males and females. A remarkable decline of resting of TP segment on electrocardiograph was also noticed. Investigations related to stress reported mental and behavioral health indications, such as depression among men and women, at follow-up after exposure via cell phones [12]. The concern about radiation exposure of RFR in children is one of the important issues. Children are affected by surrounding exposures due to cell phone radiation in a disproportionate manner. The most concerning diseases are cancer and neurological diseases. Regarding radiofrequency exposure, it was observed that children are more susceptible than adults [13].

X-ray and gamma-ray sources used in diagnosis and treatment are other sources of severe biological impairment. Effects from these exposures may vary from no consequence to death depending on the intensity and time of exposure. These exposures can be a reason for diseases such as leukemia, bone cancer, breast cancer, and lung cancer. Investigations report that X-ray exposure over the long term in CT scan can alter genetic biomaterial in cells and may cause mutations in the cell [14].

Long-term investigations reported warning regarding the imbalance of ecosystem. Intensity of radiofrequency (RF) and exposure period of non-ionizing radiation cause harm to the bio system and ecosystems also [15, 16]. Various studies have shown the ill effects of radiofrequency electromagnetic field on bees, fruit flies, frogs, birds, and bats, but the long-term studies of such exposures are inconclusive, rare, and almost non-existent in India. Short-term studies on the adverse impacts of radiofrequency electromagnetic field on frogs, honeybees, house sparrows, bats, and even humans are distressing. Honeybees emerge to be very sensitive to electromagnetic field. It has been noticed that honeybee colonies are reducing day by day. If their behavioral responses are scientifically recorded and interpreted, it could be used as an indicator of EMF pollution on biological species [17]. Einstein believed, "If the bee disappears from the surface of the earth, man would have no more than four years to live."

Field intensities of radiation by cell phone towers go down speedily with increase in distance from the cell phone tower. The attenuation of power decreases with increase in distance from the tower. The human body contains 70% of liquid (or fluid) in the body. When the human body is exposed to the EMR, it absorbs maximum radiation. The absorption is similar to cooking food in the microwave oven. The height of a human is much more than the wavelength of the transmitting frequencies of the cell phone tower. Therefore, there is possibility of many resonances in the body. Consequently it may cause restricted heating within the body. Finally, there may be dehydration in eyes. It causes eye pain, dehydration of fluid in the brain, the joints, the heart, the abdomen, etc., causing overall detrimental effects [18].

The radiation effects due to non-ionizing radiation on the human body are expected to be dangerous and never-ending at all. The main aspect to be considered for investigating detrimental effects include duration of cell phone use, number of devices used at a time, average calls received and dialed per day, position of keeping the cell phone, information about use of such devices by children, information about hypersensitive people, information about cell phone towers in the vicinity of crowded areas, radiation effects due to cell phone towers, preventive measures to be taken to reduce the radiofrequency radiation, biological effects such as fatigue, dizziness, loss of attention, memory loss, disturbance in sleep, mental weakness, and other associated health problems.

The major hazard of EMR is the emission of radiation by the permanent structure used in mobile communication. These infrastructures include base tower stations and their antennas. Due to exponential increase in cell phone manufacture and cell phone users all over the world, the numbers of base station towers or cell phone towers are also increasing considerably. Cell phone towers offer the connection between cell phone users in the process of transmission and receiving signals. Compared with mobile handsets, a cell phone tower emits radiation continually 24 h a day. The absorption of EMR affects more the people staying near the tower. Even plants and animals near the tower exhibit adverse effects on their growth due to the radiation emitted by the cell phone tower.

Field intensities of radiation by cell phone towers go down speedily with increase in distance from the base tower station (BTS).

Consequently it may cause restricted heating within the body. Finally, there may be dehydration in eyes. It causes eye pain, dehydration of fluid in the brain, the joints, the heart, the abdomen, etc. causing overall detrimental effects [18]. Maximum radiation emitted in the direction perpendicular base transmission station is indicated in the Fig. 10.1. The primary lobe has maximum radiations while the secondary lobe contains low radiations.

The existing International Committee on Non Ionizing Radiation Protection norms are entirely based on the thermal effects. Diverse epidemic and laboratory investigations had revealed the considerable biological effects far low than these standards adopted. The effects of RF radiation developed become more prominent for 8 to 10 years of exposure. These effects are not seen at the early stages of exposure because the body has definite resistance mechanisms. With increase in stress, the response is reduced and body cells receive less protection against the harm. Thus, in the long term, even low intensity could be harmful.

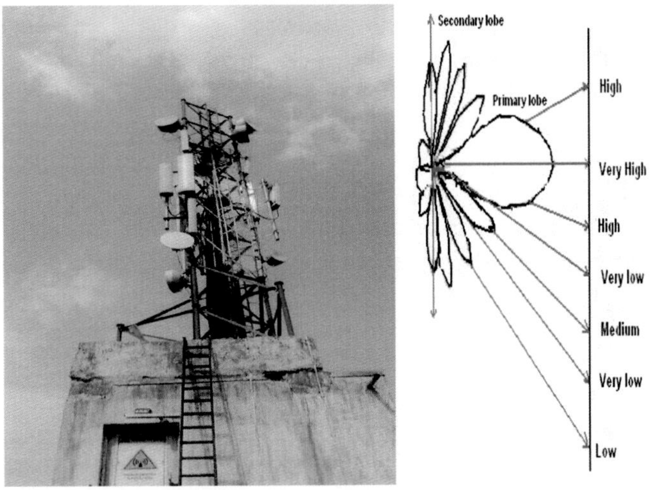

Figure 10.1 Base transmission station and radiation exposure.

Weak EMF releases Ca^+ (calcium) ions from the membrane of a cell. The seepage of calcium ions speeds up the growth and healing rapidly. However, unfortunately it also promotes the initiation of tumors. Loss of Ca^+ causes leakage in the membranes of lysosomes, which causes DNA damage. Another instance of DNA damage is enlarged free radical formation inside cells [19].

Gandhi et al. observed that children are more susceptible to radiofrequency radiation. This is because children have thin skulls, developing nervous system, and partly developed myelin cover on neurons as shown in Fig. 10.2 [20]. Both an expectant woman and her fetus are susceptible to RF radiations. MW radiation harms the placental barrier, signifying that the use of cell phone should be avoided by pregnant women.

Investigations on the RF field indicate the correlation with various diseases like Alzheimer's disease, motor neuron disease, and Parkinson's disease with the radiofrequency exposure. All diseases are concerned with the decrease of specific neurons in the body. Hence, people living near base tower stations are at a larger risk of developing neuropsychiatric problems and health problems. The complaints about headache, reminiscence loss, vomiting, giddiness, lack of sensation, stinging, changed reflexes, muscle, joint and leg/foot pain, sadness, and sleep disorder have been reported [21].

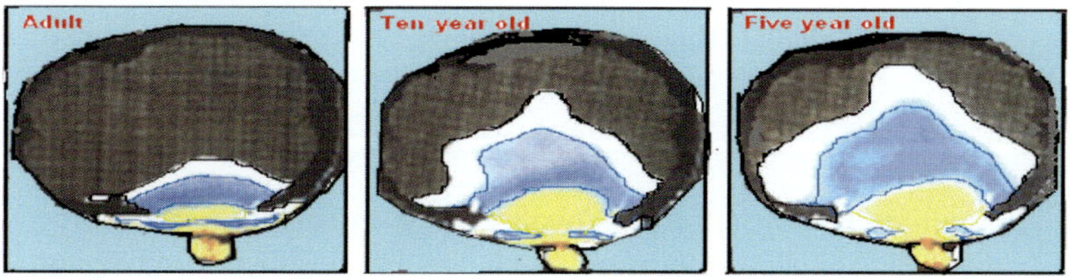

Figure 10.2 Radiation on the head of an adult, 10-year-old child, and 5-year-old child.

Other detrimental effects include epileptic fits, paralytic movements and neurosis. All effect points to the current exposure standards for microwaves, which are unsafe for long-term exposure. Different health symptoms of RF exposure faced by the inhabitants within 50 m and beyond 50 m from the tower were analyzed and compared. The results were also analyzed gender wise. It was reported that the inhabitants residing within 50 m of the location of a base tower station had more health complaints than those living beyond 50 m. It was also found that females had more health complaints than males [22].

Changes in electroencephalograph (EEG) of brain activity, neurotransmitters, metabolic changes like calcium ions, loss of concentration, slow response time in school children, increase in blood pressure in healthy men, and damage to eye cells have been reported in various studies on the effects of EMR [23].

10.2 Case Study

In this section, the effects of cell phone radiation on cardiovascular parameters are discussed.

10.2.1 Methodology

10.2.1.1 Sample selection

Experiments were carried to observe the effect of the continuous communication via cell phones on the cardiovascular system of young and healthy volunteers. Participants selected (n = 50) of age 21.8 years (±2.04) were physically fit. The volunteers on medical treatment, with occurrence of any rigorous cardiovascular syndrome together with arterial hypertension; with any pains of the neurological system; with metabolism issues that could affect the heart rate; with serious arrhythmias; and addicted to smoking and treated for alcoholism were excluded.

Volunteers with normal health were included in the study. It was ensured that volunteers included in study had been using cellular phones for more than 3 years prior to the study. After informal discussion with volunteers, it was observed that they were using cellular phones for about 9–10 h per month. It was also noticed that the average time of each cell phone call was 3 min. The volunteers averagely dialed 10 calls per day and minimum 300 calls per month. Necessary consents for the study were obtained from all students included in the study. All subjects were informed to avoid caffeinated drinks and were told not to have unnecessary activities together with physical exercises and aerobatics 12 h prior to the collection of data. They were also informed not to have food or tea prior to experiments. The volunteers were provided information regarding the equipment that would be used for examining the health parameters, procedure of study, and experimentation.

10.2.1.2 Experiments

Systolic and diastolic blood pressures were noted for all selected volunteers using the Diamond Sphygmomanometer deluxe model. The study was completed in a quiet, airy, and spacious room in sitting position. The recording of the respiratory rate (RR) and saturated oxygen percentage SpO_2 along with systolic blood pressure (SBP), diastolic (DBP), HR, and pulse rate (PR) of a volunteer was completed with the help of simple six-channel multipara monitor from Scientech Company of Model Caddo 19B, as shown in Fig. 10.3. The machine was capable of recording and displaying very important signals such as the ECG wave form, heart rate, respiratory rate, SpO_2, and pulse oximetry pattern. Four sensors (also called electrodes) combined in a cable were employed for measurement to make the unit competent for recording physical parameters as indicated in Fig. 10.4. For the measurement of BP, a cuff was used as in the measurement using the sphygmomanometer.

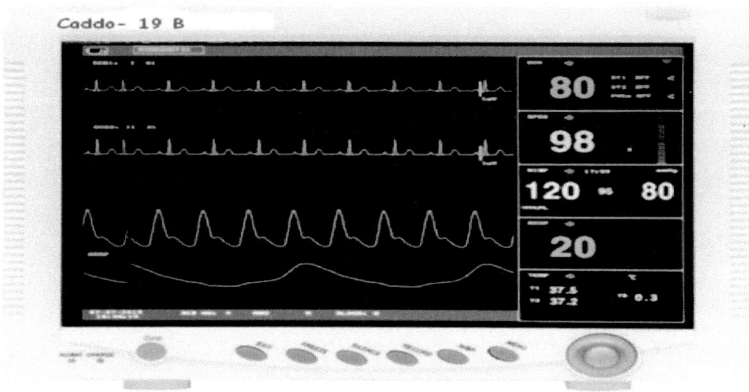

Figure 10.3 Multiparameter machine.

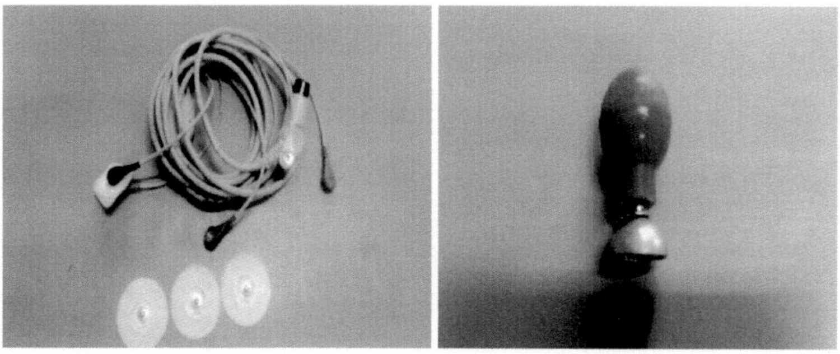

(a) Disc electrodes (b) chest electrode

Figure 10.4 Electrodes used for study.

The pilot study was carried out for continuous 1 h communication. The time of maximum exposure was determined by observing the effect of 1 h exposure. Recordings of health parameters were completed for about 20 min from 10 am to 11 am every day under the same circumstances, at the same position of the study, and in sitting position. Readings were taken prior to the mobile call (control), after 10 min with a cellular phone used continuously (exposure for 10 min), and 20 min after continuous talk on the cell phone. During the experimentation, each volunteer was receiving radiation exposure released by 1,800 MHz frequency cell phone held in the right hand.

The mobile handset used for the experiment was GSM Karbonn A1+ model, easily available on the market. The specific absorption ratio (SAR) value for model Karbonn A1+ phone was 0.633 W/kg. The cell phone was kept in the "on position" during the experiments so that the consequence of logging into the mobile networking at the maximal release of radiation could be avoided. General, non-debatable, non-exciting, or non-irritating and unbiased issues were discussed during the cell phone communication to reduce the stress linked with talking and successive excess activity of the autonomic nervous system. The volunteers were in a trouble-free sitting position and not allowed to make any bodily movement.

The experimental arrangement is shown in Fig. 10.5. These observations were taken in a quiet room at 23°C. Various parameters were recorded for the investigation when a volunteer was continuously communicating with the other person on normal issues.

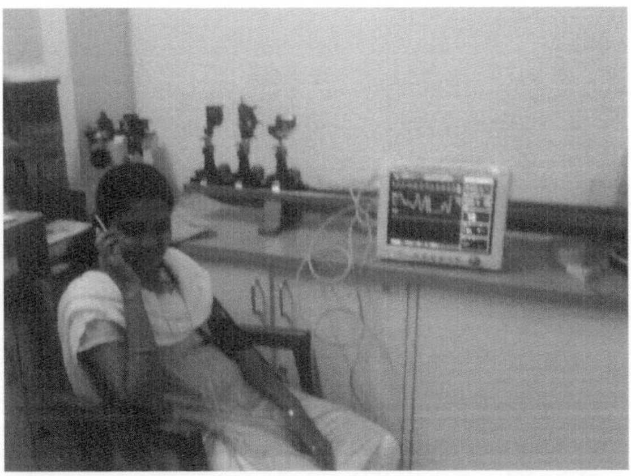

Figure 10.5 Experimental setup.

10.3 Results and Discussions

The typical variations in recorded parameters of a volunteer for 1 h communication are shown in Table 10.2.

Table 10.2 Variation in health parameters for 1 h conversion on cell phone

	% variation in the parameters for various durations		
Parameters	**20 min**	**40 min**	**60 min**
Systolic blood pressure (mm of Hg)	6.93	2.97	3.96
Diastolic blood pressure (mm of Hg)	4.23	2.80	1.39
Respiration rate (per min)	26.32	15.79	15.79
Heart rate (beats per min)	16.04	12.24	2.45
SpO_2	11.24	2.2	8.99
Pulse rate (per min)	1.3	0	5.33
Temperature	3.01	3.61	3.61

A typical variation in health parameters using Multipara Caddo 19 B during continuous talk on the cell phone for 60 min shown in Fig. 10.6.

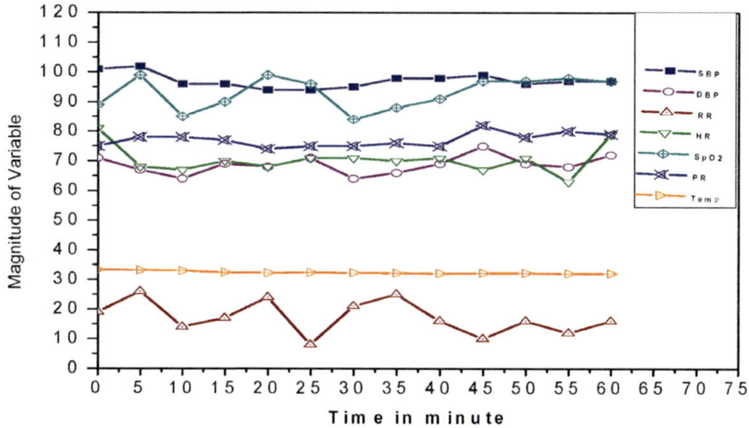

Figure 10.6 Variation in health parameters for continuous 1 h communication.

Statistical p value is the probability of finding the experimental effect under a null hypothesis. It is determined by comparing the test statistic against tabulations of specific percentage points of statistical distributions. In particular, p values less than 0.05 are often reported as "statistically significant," and interpreted as being small enough to justify rejection of the null hypothesis. The value $p = 0.05$ threshold is an arbitrary one that usually used in medical and psychological research. About 95% interval is often interpreted as indicating a range within which we can be 95% certain of the true effect.

Table 10.3 shows comparison during conversion on the cell phone with mean, standard deviation (S.D.), standard error (S.E.), and statistical p values for 10 min and 20 min of continuous talk for 50 volunteers. The analysis of data was done using online software Student's t test.

Table 10.3 Comparison of p value after cell phone exposure

Health Parameter	Exposure time	S.D.	S.E.	*p* value
SBP mm of Hg	0 min (N)	14.66	2.93	—
	10 min	18.40	3.68	0.49
	20 min	15.23	3.047	0.32
DBP mm of Hg	0 min (N)	9.25	1.85	—
	10 min	12.16	2.43	0.2885
	20 min	10.89	2.17	0.1724
RR per minute	0 min (N)	4.13	16.48	—
	10 min	4.05	5.62	0.2347
	20 min	5.12	1.12	0.4786
HR beats per minutes	0 min (N)	10.98	2.19	—
	10 min	12.35	2.47	0.0544
	20 min	13.46	2.69	0.0027
PR per minute	0 min (N)	14.78	2.95	—
	10 min	12.56	2.51	0.667
	20 min	13.89	2.77	0.5572
SpO_2 in %w	0 min (N)	3.19	0.63	—
	10 min	4.057	0.81	0.8172
	20 min	1.89	0.378	0.2636

Variation in the heart rate was on the threshold of significant after 10 min. It is more significant for exposure of cell phone communication for 20 min. It can be concluded that continuous mobile communication for 20 min affect the heart rate. For a group of 50 volunteers involved in study, the changes in the heart rate after 10 min exposure were at the threshold of significance (p = 0.0544), but it became strongly significant (p = 0.0027) after mobile communication for 20 min.

The gender-wise effect of continuous talking on the cell phone for 20 min from the calculated p value is shown in Fig. 10.7.

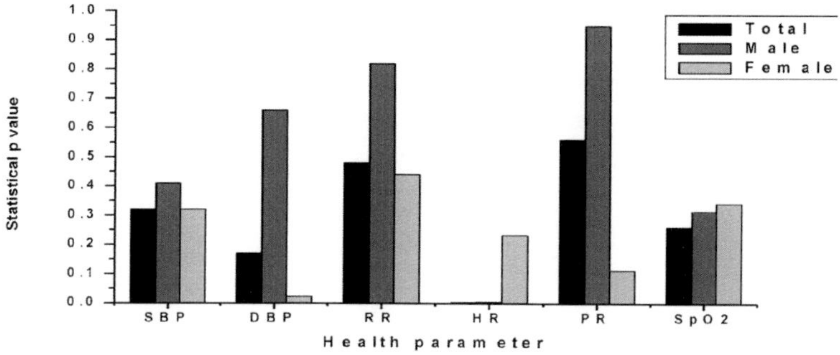

Figure 10.7 Variation in p values (gender-wise) for 20 min exposure.

10.4 Conclusions

It has been observed that p values for systolic blood pressure, respiration rate, pulse rate, and SpO_2 are not significant. These experiments were conducted in the morning from 10 am to 11 am. Other investigators conducted studies from 1.00 pm to 3.00 pm. and 9 am to 1 pm. The exposure of cell phone radiation in this experiment was for 20 min only, while in other cases it varied from 20 to 50 min [24]. The continuous conversation on the cell phone might change the rate of breathing. Therefore, variations in the respiratory rate may have an influence on the heart rate of a subject. This probability should be taken into account while the heart rate variability analysis is carried out considering respiration [25, 26]. The effect of cell phone radiation has not been fully understood by common people. Unfortunately, cell phone–manufacturing companies claim that there are no health effects due to radiation emitted by cell phones and towers.

Cell phone manufacture industry is like the cigarette manufacture industry. Cigarette companies publically agree that smoking is injurious to health and the same warning is printed on cigarette packets. Still millions of people are becoming victims of smoking, inviting pulmonary and respiratory disease like tuberculosis and asthma. We cannot view radiation emanated from cell phones or smell it, but its effects are observed after long-term exposure. In addition to cell towers, other sources such as wireless phones, computers, laptops, TV towers, FM towers, AM towers, and microwave ovens are sources of radiation. We are continuously exposed to these radiations. Firm rules should be applied to protect the public from such harmful radiations. Commercial benefits from the telecommunication industry to government are obvious and considerable, but it should be kept in mind that real wealth of a country is in the health of its people.

10.4.1 Limitations of the Study

The data collected for the survey was limited to 50 volunteers. We feel that for biological analysis, its sample size can be increased for all age groups and all parts of society. The investigators feel that large data can be used for better conclusions. Due to lack of a spectrum analyzer, it was difficult to measure the radiation exposure from the cell phone and its SAR value. The combined study of SAR values with a large number of volunteers will be helpful for further study. The present study can be used as a basis for further research in this area.

10.4.2 Recommendations

It is suggested that the radiation exposure can be reduced practically by using cell phones in emergency only, curtailing the talk time, and communicating by text whenever possible. To evade jeopardy due to EMR, simple steps can be used to reduce the exposure as follows:

(1) Talk time on the phone should be minimized. Discussion on a cellular phone for longer than 20 min increases the temperature of the ear lobe and the brain. Hence, it is suggested to avoid long conversations and frequent use.

(2) If conversion for longer period is needed, cell phone should be kept at more distance from body, especially the head, because the exposure level of radiation decreases with increase in distance of the cell phone from the body. Hence, a handset should be placed at least 15–18 cm away from the body.

(3) It is recommended not to keep the cell phone in close proximity to the body such as in a shirt pocket or on a belt clip.

(4) It is preferable to avoid use of cordless phones in the house. It is suggested to use the cell phone only when a usual telephone is unavailable. Ordinary landline phone should be used as they emit less radiation.

(5) Cell phones should be used with the lowest output power. The speaker phone should be used whenever possible. The signal intensity should be always ensured. Cell phone calls should be avoided when signal is weak because cell phone radiation is the highest at this stage.

(6) It is recommended not to speak on a cell phone inside a bus, train, and airplane. Confined areas accumulate radiation and as a result radiation exposure becomes higher in these enclosed metal surroundings. Cell phones should be used in open spaces as often as possible.

(7) The user has to remain out of Wi-Fi fields. Long time stay in Wi-Fi zones should be avoided as radiation is continuous. The wired internet connections should be preferred in schools, colleges, and educational institutes.

(8) It is recommended that children under 16 years of age be discouraged from using cell phones and playing games on them. They are more susceptible to RF radiation than adults. Pregnant women should avoid the use of cell phones, as developing child is more susceptible to RF radiation.

(9) Cell phones should be switched off when not in use. During sleep, cell phones should not be kept on or under the pillow. It releases signals in the stand-by mode, too.

(10) It has been observed that ferrite bead attached to headset decreases RF interference. Ferrite is a semi-magnetic substance created from iron oxide (rust) alloyed with other materials. It is commonly used on computer cables, to reduce or eliminate radiofrequency.

(11) For pacemaker users, necessary precautions should be taken to not keep cell phones near the pacemaker circuit. Use of a cell phone while driving should be strictly avoided.

10.4.3 Scope for Future Work

The importance of health hazards due to cell phone and cell phone towers has not been still understood by a layperson. Cellular phone companies and cellular operators claim that there are no ill effects of using cell phones. Therefore, for installing mobile/cell towers, following rules and regulations should be strictly pursued. Violations of the rules should be punishable irrespective of the income received from the company for allowing the erection of cell phone towers in residential zones.

(1) Consent from the local self-government is essential with some license fees along with the consent of the people staying near the tower. Consent from the Department of Telecommunications is obligatory with observations and visits by experts.
(2) Rules and regulations created by the TRAI have to be observed and sustained. TRAI should guarantee that quality service being provided to the customer/user. In addition, TRAI should examine and carry out the survey about the quality of service. TRAI should provide directives to improve the services and the quality of service.
(3) The number of towers to be erected should be on the basis of requirement and number of the consumers. Towers to be erected should be away from highly populated areas, schools, and hospitals.
(4) Instead of a single huge tower emitting larger radiation, erection of multiple small towers should be encouraged.

Acknowledgments

The authors are grateful to the Principal, Dr. M. G. Chaskar, for his kind support to use the facility available for completing the present work. The authors also acknowledge Dr. Popat Tambade for his valuable support.

References

1. Portier C. J., Wolfe M. S. (eds.) (1998). Assessment of health effects from exposure to powerline frequency electric and magnetic fields-Working group report, NIH Publ. No. 98–3981, Research park, NC.
2. Banik S., Bandyopadhay S., Ganguli S. (2003). Bioeffects of microwave: A brief review, *Bioresource Technol.*, 87, pp. 155–159.
3. Joel M. Moskowitz. (2012). *BioInitiative* 2012. A Rationale for biological based exposure standards for low intensity electromagnetic radiation, pp. 5–6.
4. Nittby H., Brun A., Eberhardt J., Persson B. R., Salford L. G. (2009). Increased blood-brain barrier permeability in mammalian brain 7 days after exposure to the radiation from a GSM-900 mobile phone, *Pathophysiology*, 16(2–3), pp. 103–112.
5. Zhao T., Zou Z., Knapp P. (2007). Exposure to cell phone radiations up regulates apoptosis genes in primary culture of neurons and astrocytes, *Neuro Sci. Lett.*, 412(1), pp. 34–38.
6. Wdowiak A., Wdowiak L., Wiktor H. (2007). Evaluation of the effect of using mobile phones on male fertility, *Ann. Agric. Environ. Med.*, 14, pp.169–172.
7. Agarwal A., Desai N. R., Makker K., Varghese A., Mouradi R., Sabanegh E. (2009). Effect of radiofrequency electromagnetic waves (RF-EMF) from cellular phones on human ejaculated semen: An in vitro study, *Fertil. Steril.*, 92(4), pp. 1318–1325.
8. Tuengler A., von Klitzing L. (2012). Mobile phones, electromagnetic hyper sensitivity and the precautionary principles, *Electromagn. Biol. Med.*, pp. 1–10.
9. Lahkola A., Tokala K., Auvinen A. (2006). Meta-analysis of mobile phone use and intracranial tumors, *Scand. J. Work Environ. Health*, 32(3), pp. 171–177.
10. Lahkola A., Auvinen A., Raitanen J., Schoemaker M., Helle C., Christensen H., Feychting M., Johansen C., Klaeboe L., Lonn S., Swerdlow A. J., Tynes T., Salmien T. (2007). Mobile phone use and risk of glioma in 5 North European Countries, *Int. J. Cancer*, 120, pp. 1769–1775.
11. World Health Organization (2007). ELF *Health Criterion Monograph*, Neurodegenerative Disorders, p. 187.

12. Thomas S., Henrich S., von Kries R., Radon K. (2010). Exposure to electromagnetic fields and behavioural problems in Bavarian children and adolescence, *Eur. Jr. Epidomiol.*, 25(2), pp. 135–141.

13. Sly J. L., Carpenter D. O. (2012). Special vulnerability of children to environmental exposure, *Rev. Environ. Health*, **27**(4), pp. 151–157.

14. Zamanian A., CyHardiman. (2005). Electromagnetic radiation and human health: A review of sources and effects, *High Frequency Electron.*, pp. 16–26.

15. Levitt B. B., Lai H. (2010). Biological effects from exposure to electromagnetic radiation emitted by cell tower base stations and other antenna rays, *Environ. Rev.*, 18, pp. 369–395.

16. Sinik V., Despotovic Z. (2012). Influence of electromagnetic radiation on health of people-Limit for exposure to EMF, *Infoteh-Jahorina*, 11, pp. 417–421.

17. Reshetnyak S. A., Shcheglov V. A., Blagodatskikh V. I., Gariave P. P., Maslov M. (1996). Mechanism of interaction of electromagnetic radiation with a biosystem, *Laser Phys.*, 6(4), pp. 621–653.

18. Pathak P. P., Kumar V., Vats R. P. (2003), Harmful electro-magnetic environment near transmission tower, *Indian J. Radio Space Phys.*, 32, p. 238.

19. Kumar S., Pathak P. P. (2011), Effects of electromagnetic radiation from mobile phone towers on human body, *Indian J. Radio Space Phys.*, 40, pp. 340–342.

20. Gandhi O., Lazii P., Furse C. (1996). Electromagnetic absorption in the head and neck for mobile telephones at 835 and 1900 MHz, *IEEE Trans. Microwave Theory Tech.*, 44(10), 1884–1897.

21. Abdel-Rassoul G., El-Fateh O. A., Salem M. A., Michael A., Farahat F., El-Batanouny M., Salem E. (2007). Neurobehavioral effects among inhabitants around mobile phone base stations, *Neurotoxicology*, 28(2), p. 434440.

22. Pachuau L., Pachuau Z. (2014). Study of cell tower radiation and its health hazards on human body, *IOSR J. Appl. Phys.*, 6(1–1), p. 06.

23. Sharma A., Rani B., Yadav R. K., Sharma A., Sharma S., Maheswari R. (2012). Cell phone radiation and health impacts: An overview, *Int. Jr. Geol. Earth Environ. Sci.*, 2(2), pp. 245–256.

24. Tahvanainen K., Nino J., Halonen P., Kuusela T., Laitinen T., Lansimies E. (2004), Cell phone use does not acutely affect blood pressure on heart rate of humans, *Bioelectromagnetics*, 25, pp. 73–83.

25. James R Jauchem. (1997), Exposure to extremely-low-frequency electromagnetic fields and radiofrequency radiation: Cardiovascular effects in humans, *Int. Arch. Occup. Environ. Health*, 70, pp. 9–21.
26. Bernardi L., Wdowczyk-Szule J., Valenti C., Castoldi S., Passino C., Spadacini G., Sleight P. (2000). Effects of controlled breathing, mental activity and mental stress with or without verbalization on heart rate variability, *J. Am. Coll. Cardiol.*, 35, pp. 1462–1469.

Index